PACEMAKER®

Health

GLOBE FEARON

Pearson Learning Group

Pacemaker® Health, Third Edition
We would like to thank the following educators, who provided valuable comments and suggestions during the development of this book.

Consultants

Content Consultants: **Louis J. Cantafio Jr., Ph.D.,** Ramapo College of New Jersey, Mahwah, New Jersey; **Richard Lowell, Ed.D.,** Ramapo College of New Jersey, Mahwah, New Jersey; **Susan Petro, D.V.M.,** Ramapo College of New Jersey, Mahwah, New Jersey
Educational Consultant: **Kristine M. Meurer, Ph.D.,** New Mexico Public Education Department, Santa Fe, New Mexico
ESL/ELL Consultant: **Elizabeth Jiménez,** GEMAS Consulting, Pomona, California

Reviewers

Content Reviewer: **Sean M. Devine,** Waldwick High School, Waldwick, New Jersey
Teacher Reviewers: **Janet B. Barnette,** D.R. Hill Middle School, Duncan, South Carolina; **Sherrie Coleman,** Crayton Middle School, Columbia, South Carolina; **Elizabeth McElwain,** Waldwick High School, Waldwick, New Jersey

Project Staff

Art and Design: Tricia Battipede, Evelyn Bauer, Sharon Bozek, Jane Heelan, Bernadette Hruby, Dan Trush, Heather Wendt Kemp
Editorial: Stephanie Petron Cahill, Martha Feehan, Jennie Rakos, Marilyn Sarch, Jeff Wickersty
Production and Manufacturing: Irene Belinsky, Cheryl Golding, Alia Lesser, Cathy Pawlowski
Marketing: Maureen Christensen, Doug Falk, Linda Hoffman
Publishing Operations: Kate Matracia, Debi Schlott
Technology: Joanne Saito

About the Cover

Health is the study of the wellness of people. Overall health consists of physical, emotional, and social health. Physical health involves the wellness of the body. Emotional health deals with feelings, attitudes, and behaviors. Social health is the wellness of a person within society. The images on the cover of this book represent these three parts of health. The no smoking symbol, the cardiogram, and the fruits and vegetables all represent physical health. We take care of our bodies when we exercise, eat healthy foods, and do not smoke or do other harmful activities. The group of smiling young people in a huddle represents emotional health with the positive feelings shown and social health in getting along with others.

ISBN 0-13-024692-1

Printed in the United States of America

2 3 4 5 6 7 8 9 10   08 07 06 05

Globe Fearon
Pearson Learning Group

1-800-321-3106
www.pearsonlearning.com

Contents

A Note to the Student

Your grandparents and their parents and grandparents faced different health problems than you do. Just 100 years ago, doctors could not do much for people who were sick with pneumonia, smallpox, or the flu. Most people died of any infectious diseases they caught. People who had mental disorders had little hope of effective treatment.

Advances in health science have changed that. Of course, people still suffer from diseases related to aging. However, vaccination programs and antibiotic drugs prevent many diseases and control others. New drugs have also greatly improved the care of people with mental disorders. Today's health hazards have a lot to do with the choices people make. Unhealthy habits and environmental pollution are the greatest health concerns today.

Learning about health is a basic part of your education. This book will give you the information you need to take good care of yourself. You will learn about the human body and how it functions. You will find out what to do about common health problems. You will learn why proper diet, exercise, and rest are important to your health. You will also learn about good emotional health habits that will help you have a happier, less stressful life.

Throughout the book you will find notes in the margins, or side columns, of the pages. Sometimes they will give you information about material that you are learning. Often they will provide fascinating health facts. A few will remind you of something you already know.

There are several study aids in the book. The beginning of every chapter contains a list of **Learning Objectives**. This list can help you focus on the important points in the chapter. A **Words to Know** list gives you some health vocabulary you will find in your reading. Each chapter also has a **Life Skill** page that relates health information from the chapter to real-life situations. At the end of each chapter, a **Summary** will give a review of what you just learned.

In order to help you with **Online Health Projects**, there are boxed instructions for using the HealthLinks® Web site in each chapter of this book. This site is designed to help you find more information for reports and other research projects. Whenever you see this box, log on to www.scilinks.org/health and use the key terms and codes provided to start your search. Your teacher will help you register.

We hope you enjoy reading and learning about modern health science. Everyone who put this book together has worked hard to make it useful and interesting. The rest is up to you. We wish you well in your studies. Your success is our greatest accomplishment.

Skills for Healthy Living Handbook

What Is Health?

Factors That Influence Health

Life Skills

- Making Healthy Choices

- Setting Goals

- Accessing Valid Health Information

- Communicating and Advocating for Others

- Analyzing a Personal Health Assessment

- Getting Help With Your Health

- Understanding Health Insurance

Before You Read

In this Handbook, you will learn about the meaning of overall health. You will learn about things around you that can affect your health. You will also learn some important life skills that can help you to stay healthy.

Before you read, ask yourself the following questions:

1. What do I already know about overall health?

2. What questions do I have about things that can influence my health?

3. Do I use any life skills now that I know help me to stay healthy?

These healthy people are making the most out of what they have.

SKILLS FOR HEALTHY LIVING HANDBOOK

What Is Health?

Can you know if a person is healthy simply by looking at him or her? Sometimes people who appear to be healthy are not, and sometimes people who seem to be unhealthy are healthy in ways that cannot be seen. Both your mind and your body affect your overall health. A healthy person is someone who makes the most out of the mind and body he or she has.

Often when we think of health, we think first of physical health. **Physical health** refers to the wellness of the body. You can improve your physical health by taking good care of yourself.

Emotional health has to do with feelings, attitudes, and behavior. How you handle problems determines your emotional health. Most emotionally healthy people can work their way through problems in a constructive way.

Social health is the ability to get along with others. It involves your ability to communicate, to be a good friend, and to resolve conflicts.

Physical, emotional, and social health are closely related. Sometimes when the mind has too many problems, the body is affected. Too much stress can contribute to physical illnesses, such as headaches and heart disease. It also affects the way you relate to people. If you are feeling a lot of stress, you may become angry more easily. Physical, emotional, and social health combine to make up what is called your **overall health**.

Physical Health | Social Health

Emotional Health

OVERALL HEALTH

Factors That Influence Health

Many things affect your health. Heredity is one important factor. **Heredity** is the passing of certain traits from parents to their children. Healthy, long-living parents often have healthy, long-living children. But sometimes parents can pass harmful traits, such as diseases, to their children.

The environment also plays a role in your health. Your **environment** is everything that makes up the world around you. A person who lives in a smoke-filled environment day after day is more likely to get lung disease. A person who lives with someone who loves him or her is more likely to have good emotional health.

You are the most important factor in your overall health. You make choices about your health all the time. Eating and exercise habits, drug use and smoking—all of these things figure into your health. The first step toward achieving good health is taking responsibility for your health.

▶**CRITICAL THINKING** Describe how a person in a wheelchair might be in better health than a star athlete.

Your exercise habits are one important factor in your overall health.

Life Skills

You have many health choices to make today. You will have many more health choices in the years ahead. Sometimes it is not easy to know which choices to make. Making the right choice can keep you healthy and even save your life.

Making wise choices is just one skill that will help you to maintain good health. There are many other skills that will lead to good health. By understanding and applying these skills, you can improve your own health and the health of the people around you.

Making wise choices when food shopping will help maintain good health.

Skills for Maintaining Good Health

- Making Healthy Choices
- Setting Goals
- Accessing Valid Health Information
- Communicating and Advocating for Others
- Analyzing a Personal Health Assessment
- Getting Help With Your Health
- Understanding Health Insurance

CRITICAL THINKING Which of the skills listed above do you already know how to do? Which ones are new to you? How do you think these skills will help keep you healthy?

Making Healthy Choices

You make decisions about your health every day. As a teenager, you are faced with choices that can seriously affect your health and your life. It is important to think carefully about your choices so you can make the best decision. A well-thought-out process can help you make healthy choices. This process is often called the **decision-making model.** There are seven steps in the decision-making model. These steps are listed below:

STEP 1 Identify the decision you need to make.

STEP 2 List as many choices for your decision as possible. If you have many choices, you will have a better chance of making a wise decision.

STEP 3 Cross out choices that are harmful or go against your beliefs.

STEP 4 Read each remaining choice. Think about what might happen if you select it.

STEP 5 Select the choice that will probably have the best results for you and others.

STEP 6 Carry out your choice.

STEP 7 Think about the results of your decision. That way, you will know whether to select that choice the next time.

The U.S. Centers for Disease Control and Prevention finds that most major health problems are caused by six types of risk behavior. **Risk behaviors** are choices that put your health or the health of others at risk.

▶ CRITICAL THINKING Choose one of the risk behaviors from the list at the right. Apply the steps of the decision-making model to doing that behavior.

RISK BEHAVIORS

- using alcohol and other drugs
- using tobacco
- engaging in violent behavior
- eating a poor diet
- not getting enough exercise
- practicing risky sexual behavior

Source: Centers for Disease Control and Prevention

Setting Goals

A **goal** is something you want to do. Goals can give you direction. If you want to improve some aspect of your health, setting goals can help. For example, you may think a lot about getting into better physical condition, but sometimes it is hard to get started. Setting goals can help you become more physically fit.

To be helpful, a goal must be specific, realistic, and important to you. Many long-term goals can be broken into smaller short-term goals. After you have set a goal, you need to make an action plan. An **action plan** is a list of steps that will help you reach your goal. The following are steps for making an action plan:

Having and keeping a part-time job is a great goal.

STEP 1 Write down your goal. Write your goal in your journal or diary, or write it on a sheet of paper. Tape the paper up where you will see it often.

STEP 2 List steps to reach your goal. Break a long-term goal into several smaller goals. Break smaller goals into steps.

STEP 3 Set up a timeline. Decide how much time each step should take. Then, write each step on a chart or calendar. Keep track of the steps on your chart.

STEP 4 Identify any obstacles. List things that might get in the way of your goal. Write ways you might overcome them.

STEP 5 Identify sources of help. Decide who or what could help you reach your goal. You and a friend might help each other work toward the same goal.

STEP 6 Check your progress. As you work toward your goal, review your plan. Check whether you are on schedule. If your plan is not realistic, make changes so you can still reach your goal.

Accessing Valid Health Information

Today, we have more information available to us than ever before. We need accurate health information to be wise healthcare consumers. Many advertisements, articles, and Web sites provide good information about health problems and ways to improve your health. But, you cannot believe everything you read. Some health products seem too good to be true, and others might be dangerous. How do you know what to trust? It is not always easy to know. Following the guidelines below can help you find and assess health information.

1. **Read labels.** Identify ingredients and consider possible side effects.

2. **Pay attention to the source.** Be wary of information designed to sell you something. Good sources of health information include

   • The U.S. government

   • Universities or medical schools

   • Hospitals

   • Nonprofit health organizations

   • Medical and science journals

3. **Compare different resources on the same topic.** Check other Web sites or articles. Ask your doctor's opinion.

4. **Ask yourself if the advice seems to make sense.** If something seems too good to be true, it probably is.

▶ **CRITICAL THINKING** An ad promises that a vitamin drink will help you lose 30 pounds in 2 weeks. How might you decide if the information is valid?

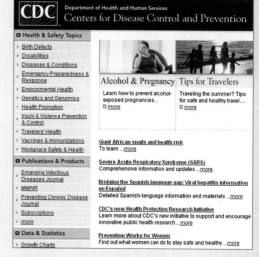

One good source of health information is the Centers for Disease Control Web site at http://www.cdc.gov/.

Communicating and Advocating for Others

Communicating

Good communication can benefit your emotional and social health. People who communicate well have better relationships with people and feel good about themselves. Poor communication can lead to stress and anger. Being a good communicator takes practice. To communicate effectively, keep the following things in mind:

- Use "I" statements instead of "you" statements. Say what you want as clearly as possible.

- Tell the person exactly what behaviors you do and do not like. Try not to criticize or blame the other person.

- Be a good listener. If you do not understand what the other person is saying, ask for an explanation. Try to summarize the other person's point of view. Consider the feelings of the other person.

Advocating for Better Health

You are responsible for more than your own good health. You also play a role in the health of your family and your community. There are many ways that you can help the health of your community. Many health organizations depend on volunteers to donate blood and distribute healthy meals to the elderly. On your own or with a group, you can encourage others to make positive health choices. You can also express your opinions about issues that affect the health of your community, such as pollution.

You can help the health of your community by running or walking in a charity race such as Race for the Cure.

▶ **CRITICAL THINKING** Identify two things you could do to improve the health of your community. Be specific.

Analyzing a Personal Health Assessment

Knowledge is the first step toward a healthier life. How healthy are you? What health risks do you face? A **personal health assessment** can help you find out. It asks you questions about your health and your knowledge of health risks.

The Classroom Resource Binder has a self-assessment page for each unit. Take each self-assessment twice. First, take it before reading the unit. For each question or statement, choose the answer that best applies to you. Each answer has a score. Add up the scores to get your total. After figuring your score, you can identify your goals for learning in the unit and improving your health. Then, take the assessment again after completing the unit. Notice if your total score has changed. Finally, record the main ideas that you learned from the unit.

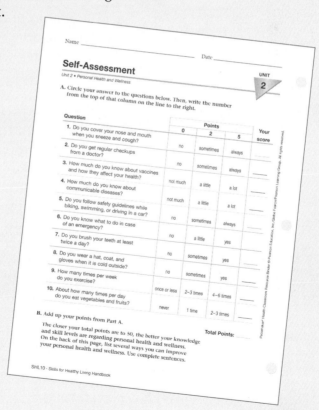

Getting Help With Your Health

Sources of Help

Sometimes you may feel that something is not quite right with your health. Maybe you are physically sick. Perhaps you feel sad or angry all the time. In that case, you might need help dealing with your emotions. If you have a health question or problem, it is a good idea to talk about it first with your parent or guardian. A school nurse or guidance counselor can also help you. The table below lists some other sources of help for your health.

Sources of Help for Good Health	
Source	**How They Can Help You**
Family doctors	They take care of most basic health needs.
Specialists	They treat only certain illnesses.
Dentists	They care for the health of teeth and gums.
Psychiatrists or psychologists	They treat mental and emotional problems.
Hospitals	They provide a wide variety of healthcare service; are always open and can handle emergencies; have rooms where patients can stay overnight.
Health Clinics	They provide low-cost healthcare.

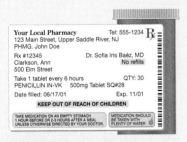

Always follow your doctor's orders when taking prescription medication.

Medications

Many illnesses can be cured or controlled with medicines. If you are sick or hurt, your doctor may write a **prescription**, which is an order for medicine. People can buy some types of medicines without a doctor's order, such as aspirin and cough syrup. They are called **over the counter medicines,** or OTCs.

Medicines come with directions on the label. The directions explain how much and how often to take the medicine. Medicines also have warning labels that tell you about possible side effects. If you have any questions, you should talk to your doctor or pharmacist.

Understanding Health Insurance

Most people do not have enough money to pay huge medical bills. Many families have **health insurance**. Health insurance is a plan that helps you pay doctor and hospital bills. People without health insurance may not get the care they need because they cannot afford it.

Some people get **group health insurance** through their job. Buying your own plan often costs much more. Health insurance plans can differ. In some plans, you pay a deductible and co-insurance. A **deductible** is a yearly amount you must spend before the insurance pays anything. **Co-insurance** is the amount you pay after the insurance pays. Other types of plans require a **co-payment.** This is a fixed dollar amount for a healthcare service, such as a doctor's visit. Some plans let you see any doctor. Other plans have a list, or **network**, of selected doctors. The chart below compares some common types of health insurance.

Health Choice USA			
NAME		COPAYS: OFFICE	10.00
Michaels, Bradley		OT/PT/SPEECH	10.00
MEMBER NUMBER		ER/URG CARE CENTER	25/15
734 153469524		EYE/EAR	15.00
BIRTH DATE EFFECTIVE CD		INPATIENT	00%
08-26-70 01-01-01 01		MENTAL HEALTH	20.00
PRIMARY CARE PHYSICIAN		ALLERGY	20%
Janis E. Withrop, M.D.		RX B	14.00
		G	8.00
Fully Insured	SEE REVERSE	Health Choice, Inc.	

Your health insurance company may send you an insurance card as proof of coverage. Take this card with you when you go to a doctor or dentist.

Health Insurance Plans			
Kind of Plan	**Advantages**	**Disadvantages**	**Cost to Individual**
Traditional	• can choose any doctor • can go to any hospital	• often lots of paperwork • little or no preventive care • yearly deductible	• highest premiums • pays 80%–90% costs
Managed Care Plans	• may choose any doctor or hospital • some preventive care covered	• less coverage for out-of-network doctors	• medium premiums • pays 60%–70% out-of-network
HMO (Health Maintenance Organization)	• most preventive care covered • very little paperwork	• only pays for certain doctors and hospitals	• lower premiums • low co-payments

CRITICAL THINKING Why do you think healthcare and prescription drugs are so important to older people?

Review

Comprehension Check

On a separate sheet of paper, write how each of the things below will benefit your health.

1. following an action plan
2. using the decision-making model
3. completing a personal health assessment
4. accessing valid health information
5. having good health insurance

Analyzing Cause and Effect

Write a sentence or two explaining what might cause the following. The information you learned in this Handbook will help you.

6. A dieter fails at his or her weight loss plan.
7. A student is suspended from school for smoking.
8. A woman finds out she is allergic to a medicine before taking it.
9. A man goes to a doctor and pays only $10.
10. A teenager argues with his parents and is grounded.

Writing an Essay

Answer the questions below on a separate sheet of paper. Use complete sentences.

11. Explain how physical, emotional, and social health affect your overall health.
12. What is the difference between traditional and HMO health insurance?
13. Name three sources of help for your health.
14. How does advocating for others help make society and your health better?
15. What are the benefits of taking a personal health assessment?

Managing Your Health

List the factors that you think influence your health. Decide whether they are positive or negative. How can you change the negative influences into positive ones?

Unit 1

The Human Body

Before You Read

In this unit, you will learn about the parts of the human body. You also will learn how all of the parts work together to help you live.

Before you read, ask yourself the following questions:

1. What do I already know about my body systems?

2. What questions do I have about how my body works?

3. Why might knowing about my body be important to staying healthy?

Athletes have greater bone mass and muscle mass than people who do not exercise regularly.

Learning Objectives

- Describe the parts that make up human body systems.

- List the parts of the skeletal and muscular systems and describe their functions.

- Name some common disorders of the skeletal and muscular systems, their symptoms, and their treatments.

- Describe some ways to keep your skeletal and muscular systems healthy.

- **LIFE SKILL:** Explain the steps you can use to treat a pulled muscle.

Chapter 1

The Skeletal and Muscular Systems

Words to Know

marrow	the soft tissue inside bones that makes blood cells
joint	a place in the body where two or more bones meet
ligament	a type of tissue that holds bones together at a joint
fracture	a break or crack in a bone
osteoporosis	a condition that makes bones weak and brittle
arthritis	a condition that causes stiff and swollen joints
tendon	a type of tissue that connects muscles to bones or other muscles
voluntary muscle	a muscle that moves with your control
involuntary muscle	a muscle that moves without your control
cardiac muscle	the muscle tissue that makes up the heart

Human Body Systems

The human body is made up of very small structures called cells. *Cells* are the basic units of structure and function in living things. Your body contains millions of cells. Groups of cells that work together make up *tissues*. Tissues that work together make up *organs*. Groups of organs that work together make up *body systems*. These body systems work together to keep your body working. For example, the skeletal and muscular systems work together to allow you to stand up straight and to move.

Talk About It

What do you know about your body systems? Name as many systems as you can.

The Skeletal System

Bones have five jobs:

1. They support your body and give your body shape.

2. They protect many organs inside your body. For example, your skull protects your brain.

3. Many bones work with muscles to help your body move.

4. Some bones have soft tissue called **marrow** inside them. This tissue makes blood cells.

5. Bones store minerals such as calcium and phosphorus.

Health Fact

You have 206 bones in your body. The longest bones in your body are your thigh bones, or femurs. The smallest bones are found in your ears.

Talk About It

Look at the drawing. Which bones are you familiar with? What are their functions?

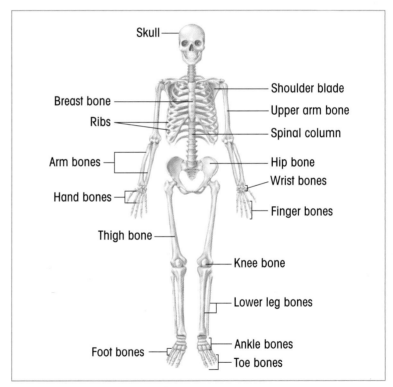

The human skeletal system makes a framework that gives the body its shape.

What Bones Are Made Of

Bones are made of both living and nonliving materials. The living materials are bone cells, blood cells, and nerve cells. These cells make up about one-third of bones. About two-thirds of bones are made up of nonliving materials, including the minerals calcium and phosphorous.

The skeleton of an unborn baby is made of cartilage. This tissue is tough but flexible. As an unborn baby develops, most of its cartilage skeleton changes to hard bone. The cartilage that is left has many functions. For example, it prevents the ends of bones from rubbing against each other.

The Structure of Bone

A bone is covered with a strong membrane. This membrane has many blood vessels that pass through it. Blood supplies living bone cells with nutrients and oxygen.

Under the membrane is compact bone. *Compact bone* is made up of living bone cells. It is also made up of the mineral calcium. Calcium in compact bone makes it hard and gives it strength. A network of tubes also runs through compact bone. These tubes contain blood vessels and nerves.

The centers of flat bones and the ends of long bones are not as hard as compact bones. These softer parts of bone are called *spongy bone*. They even look like sponges because they are filled with many spaces. The spaces are filled with connective tissue and blood vessels. Although they are softer than compact bones, spongy bones are quite strong. The spaces in them form a support structure that is lightweight. Spongy bones provide support to areas in bones where there is great pressure and stress.

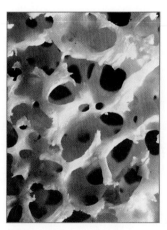

Spongy bone is filled with many spaces.

The centers of your bones are filled with marrow. **Marrow** makes blood cells. Marrow is red or yellow in color. Red marrow actively makes red blood cells. Yellow marrow is not active. It is found in the center of long bones. Yellow marrow is made mostly of fat cells.

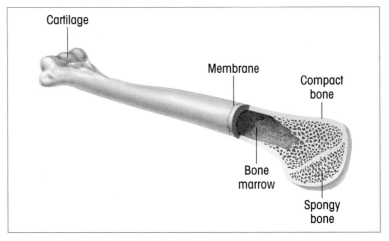

The center of a long bone can contain red marrow and yellow marrow.

Bones Change

Bones grow and change. Some cells help build bone tissue. Some cells break down bone tissue. When bone tissue is broken down, some of the minerals inside bones are released into the body for use. However, your body normally replaces those minerals with minerals from the foods you eat. The building up and breaking down of your bones changes the size and shape of your bones as you age.

At some point in your life, you may go through "an awkward stage" because of the way your bones grow. The first bones to grow quickly are usually your foot bones. Then, your arm and hand bones grow quickly. Because of this uneven growth, arms, hands, and feet may look too big for the rest of the body. This does not last long. After a while, the rest of the body catches up.

Joints

The places where two or more bones meet are called **joints**. Bones are held together by strong bands of tissue called **ligaments**.

Immovable Joints

There are several kinds of joints in your body. The joints in the skull are an example of *immovable joints*. The curved bones in a skull join together in what look like cracks. These bones do not move.

Movable Joints

The joint between your skull and neck is a *pivotal joint*. This type of joint allows your head to nod up and down. It also allows your head to move from side to side.

Your hip and shoulder joints are examples of *ball-and-socket joints*. These joints allow bones to move in several directions. *Gliding joints* allow bones to move in many directions.

The joint at your elbow is a *hinge joint*. It can only move in two directions—back and forth. Your knees, toes, and fingers also have hinge joints.

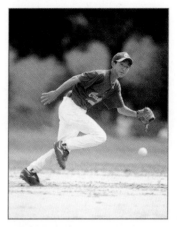

Playing sports uses different kinds of joints.

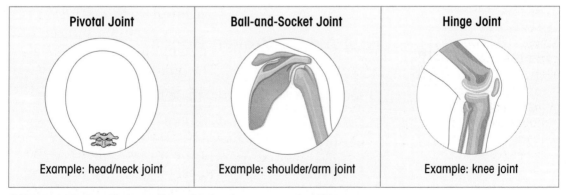

Pivotal Joint	Ball-and-Socket Joint	Hinge Joint
Example: head/neck joint	Example: shoulder/arm joint	Example: knee joint

Different kinds of joints allow the body to move in different ways.

Bone and Joint Problems

Have you ever had a broken bone? Broken bones are also called **fractures**. There are two kinds of fractures: closed and open. A *closed fracture* means that the bone is broken in place. In an *open fracture*, the skin and the muscle around the broken bone are torn.

Most fractures will heal if they are set in a cast. The cast holds the broken bone in place so the bone can grow back together. In very bad breaks, pins are sometimes placed in the bones to help the bone heal.

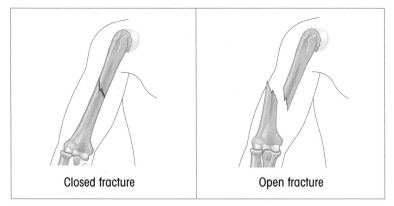

Closed fracture   Open fracture

In a closed fracture, the skin is not torn open. In an open fracture, the broken bone breaks through the muscle and the skin.

First Aid for Broken Bones

If a friend breaks a bone, he or she will need a doctor as soon as possible. Follow these first aid steps:

1. Do not move the broken bone. Do not move the person if you can help it.

2. Keep the person warm.

3. If there is bleeding, cover the broken skin with a clean bandage. If you see a bone outside the skin, do not try to force it back inside. Wait for a doctor.

4. Get help as quickly as you can.

Osteoporosis

Osteoporosis is a condition that makes bones weak and brittle. This condition is due to not having enough calcium in the bones. The bones tend to shrink and are easily broken. People with osteoporosis often have backs that are bent. This is because their bones have grown weak with age and their skeletons have become bent.

Osteoporosis has several different causes:

- One main cause is a diet that is low in calcium. Over time, the body takes the calcium it needs from the bones. The bones are weakened by the loss of calcium.

- Another cause is a decrease in certain hormones as people get older. This may be why osteoporosis usually affects middle-aged and elderly people. It is a more common problem among women than men.

To lower your risk of getting osteoporosis, make sure you are getting enough calcium in your diet. Dairy products, such as milk and cheese, contain a lot of calcium. Studies show that getting calcium in the preteen and teen years is especially important for avoiding osteoporosis. Weight-bearing exercise can also help to make your bones stronger.

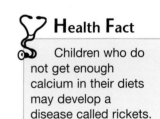 **Health Fact**

Children who do not get enough calcium in their diets may develop a disease called rickets.

Arthritis

Arthritis is a disease that causes the joints to stiffen and swell. People with arthritis may feel pain whenever they make the simplest movements, like walking down stairs or dialing a phone.

There are two kinds of arthritis. One kind affects mostly older people. It is caused by the wearing away of cartilage in the joints. The second kind attacks both young and old. The body attacks the tissues in the joint, causing them to swell. Doctors do not know exactly what causes this second kind of arthritis.

People with arthritis can treat their swollen, stiff joints with medicine. One of these medicines is called *cortisone*. Aspirin can also be used to stop pain and reduce swelling. There is no known cure for either kind of arthritis.

Keeping Your Bones and Joints Healthy

Health Fact

Studies show that a poor diet in the teen years can actually stunt your growth.

Taking care of your bones begins with safety. Car accidents, falls, and carelessness are all causes of broken bones. Always wear your seat belts in a car.

Make sure you get plenty of calcium in your diet. Milk, a high source of calcium, will keep your bones in good shape. Vitamin D, also found in milk, will help, too. A healthy diet will keep your bones growing at a healthy rate.

It is important to exercise at least three times each week. Exercise is good for your bones as well as your muscles. To protect your bones during sports, it is important to wear helmets and padding.

Write your answers in complete sentences. Use a separate sheet of paper.

1. What are four kinds of joints?

2. How can you reduce the risk of broken bones?

3. CRITICAL THINKING: Why is it important for teenagers to include the right amount of calcium in their diets?

Health and Technology

ARTIFICIAL JOINTS

Nearly one-and-a-half million people have joint replacements each year. Doctors use an artificial joint to replace the damaged joints. Most common are hip and knee replacements. Artificial joints generally wear out within ten years.

Bone Cement Fixation is one method used for joint replacement. The artificial joint is placed between two bones. Bone cement sets it in place. The new joint is designed to function like the natural joint. In another method, bone cement is not used. The artificial joint and the bone are designed to fit and lock together exactly.

Artificial joints may not be needed in the future. Recently, engineers at the University of California have developed an artificial tissue that works like real cartilage. This tissue may help repair injured or arthritic joints. The artificial tissue may also be able to bring back flexibility to people with damaged joints.

CRITICAL THINKING Is joint replacement permanent? Explain your answer.

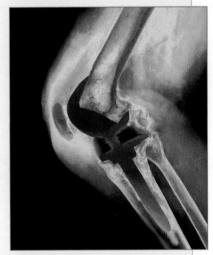

Artificial joints, such as the knee joint shown in red, replace damaged joints.

The Muscular System

Your bones cannot move on their own. They move because they are attached to muscles. **Tendons** are a type of tissue that connect muscles to bones or to other muscles. Your body has more than 600 muscles. Almost half your body weight is muscle weight.

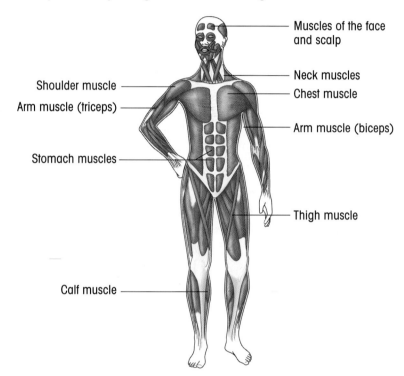

This photograph was taken using a microscope. It shows muscle tissue.

Muscles of the face and scalp

Neck muscles

Shoulder muscle

Chest muscle

Arm muscle (triceps)

Arm muscle (biceps)

Stomach muscles

Thigh muscle

Calf muscle

The human muscular system has more than 600 muscles.

Different Kinds of Muscle Tissue

There are three kinds of muscle tissue in your body: skeletal muscle, smooth muscle, and cardiac muscle. *Skeletal muscles* are attached to your bones. They are **voluntary muscles.** You can control when voluntary muscles move. This type of muscle moves your skeleton by pulling on tendons attached to bones.

Smooth muscles are **involuntary muscles**. They work whether you want them to or not. They work even while you are sleeping. Smooth muscle is found in the walls of your blood vessels and intestines. Smooth muscles are not attached to bones.

Cardiac muscle is a type of involuntary muscle found in your heart. Cardiac muscle works all the time. It keeps your heart beating.

How Muscles Work

When a muscle moves, it contracts, or shortens. When a muscle relaxes, it lengthens. Muscles can only pull. They cannot push. For this reason, skeletal muscles work in pairs.

Most voluntary muscles work in teams of two to move bones. For example, the upper arm has the biceps and triceps muscles. The biceps muscle is the muscle in the front of your upper arm. The triceps muscle is in the back of your upper arm. The two muscles work together when you move your arm.

Talk About It

Move your upper arm up and down while your other hand is on your biceps muscle. You can feel your biceps contracting, or shortening. What other muscle pairs work in the same way?

Triceps shortens — Biceps lengthens

Triceps straightens limb

Triceps lengthens — Biceps shortens

Biceps bends limb at joint

The triceps and biceps muscles balance each other. This balance allows you to move smoothly.

Sports and Muscles

Always stretch before and after exercising.

Muscles and tendons can be injured when you play sports. An injured tendon can take months, and sometimes even years, to heal.

You can avoid this trouble by doing a lot of stretching before and after exercising. Stretching keeps both muscles and tendons flexible. Good technique will also help you avoid injury. For example, lifting weights the wrong way often leads to an injury. Most important, if something hurts, stop! Do not act as if an injury is not there. You will just make it worse. If you feel pain or discomfort in your muscles or tendons, rest for a few days. If it does not go away, see a doctor.

Think About It

IS STEROIDS TESTING IN SPORTS FAIR?
Certain kinds of steroids are drugs that build muscles and strength. Steroids are actually hormones, or chemicals made by the body. Women who use body-building steroids often look more masculine. In both men and women, artificial steroids can cause skin and stomach problems, and hurt the liver.

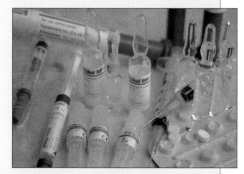

Prescription steroids come in pill and injectable forms.

Some professional athletes use steroids even though steroids are illegal. They say that almost everyone uses them anyway. They claim there is no way to win if you do not use steroids. Athletes who do not use steroids think that the drugs give the user an unfair advantage.

Before most important sports events, such as the Olympic games, athletes are tested for steroid use. If an athlete is found to have used a steroid, he or she is not allowed to compete. However, some sports do not test their players for steroid use. Many of the athletes of these sports are against drug testing.

YOU DECIDE Should some sports and not others include steroid testing? Is this fair? Support your opinion with examples.

Problems With Muscles

Suppose you lost control of your muscles little by little. That is what happens to people with muscular dystrophy, also called MD. MD is a disease that causes the muscles to waste away over time. It is a rare disease. Most children that have MD end up in wheelchairs or on crutches. At this time, doctors do not know what causes MD, but they are working on treatments for it.

The most common muscle problem is sore muscles. The soreness is caused by muscles not getting enough oxygen during exercise. Also, muscles have to get rid of wastes when exercising. By exercising regularly, your body learns to pump enough oxygen to your muscles. In time, the soreness disappears. Stretching after exercising, or "cooling down," will help lessen the pain of sore muscles.

Muscle strains are also common. A muscle is strained when it has been worked too hard. The muscle tissues are stretched and sometimes torn. To treat a strained muscle, get plenty of rest. Apply ice first. After 48 hours, put warm, wet towels on the muscle every chance you get.

A torn, bleeding muscle will cause a bruise, or a black and blue mark on the skin. A cold, wet towel can sometimes help bruises look and feel better.

✓ Check Your Understanding

Write your answers in complete sentences. Use a separate sheet of paper.

1. What are three kinds of muscle tissue?

2. What are artificial steroids?

3. CRITICAL THINKING How can you avoid a muscle injury?

Back Muscles

Back injuries are common and very painful. You can strain or tear your back muscles if you are not careful. These are some ways to avoid a back injury:

- Whenever possible, lift or carry heavy objects with the help of friends.
- Never lift a heavy object quickly.
- Stand close to the object before lifting it.
- Bend your knees, and put the weight on your legs as you lift slowly.
- Do not lift the object too high.

How To Have Healthy Muscles

Getting exercise is the best thing you can do to keep your muscles healthy. Exercise pumps blood and oxygen all over your body making you feel alert and healthy. Exercise also makes your muscles strong and firm.

A healthy exercise program begins and ends with stretching. Research shows that lack of exercise causes teenagers to gain more weight than overeating does.

Nutrition is also important for keeping muscles healthy. You should eat a variety of healthy foods. This means including meat or beans, fruits and vegetables, and breads and cereal in your diet.

✓ Check Your Understanding

Write a paragraph that summarizes the ways you can prevent back injury and keep your muscles healthy. Use complete sentences. Write your answer on a separate sheet of paper.

LIFE SKILL
Managing a Pulled Muscle

One important health skill is managing minor illnesses or injuries on your own. A common injury to muscles is a *pulled muscle*. A pulled muscle is actually small tears in the muscle. These tears cause pain and swelling. The amount of pain depends on the amount the muscle has torn. Some pulled muscles must be treated by a doctor. If you have a pulled muscle that causes severe pain and you cannot move the muscle at all, see a doctor immediately. Most pulled muscles can be treated at home. There are steps you can follow to manage a pulled muscle on your own.

First, you must take steps to limit swelling. You also need to protect the injured muscle. An easy way to remember these steps is to think of R.I.C.E. The chart below explains R.I.C.E.

The R.I.C.E Treatment Method			
Letter	Treatment	Application	Effect
R	Rest	Immediately	Prevents more damage; gives the body energy to heal
I	Ice	Apply ice or cold packs. Leave on 15–20 minutes, then off 20 minutes.	Limits swelling
C	Compression	Wrap an elastic bandage over the area.	Limits swelling
E	Elevation	Raise the injury above the level of the heart.	Reduces swelling

Answer the questions on a separate sheet of paper. Use complete sentences.

1. Why should you rest after an injury?

2. Why should you put ice on a pulled muscle?

3. What does the *E* in R.I.C.E. stand for?

Applying the Skill

After a day or two of R.I.C.E., many injuries begin to heal. What would you do if the pain and swelling did not decrease after two days?

Summary

The skeletal system is made up of bones and joints. Bones have five jobs:
1) they support the body and give it shape; 2) they protect organs inside the
body; 3) they work with muscles to help the body move; 4) some bones help
make blood cells; and 5) they store some minerals that the body needs.

Some problems with the skeletal system include fractures, osteoporosis, and
arthritis. You can keep your skeletal system healthy by practicing safe behavior,
exercising, and eating a balanced diet.

The muscular system is made up of muscles and tendons. There are three kinds
of muscles: skeletal muscles, smooth muscles, and cardiac muscles. Tendons
connect muscles to bones and other muscles.

The most common muscle problem is sore muscles. Sore muscles can be
prevented by exercising regularly. Muscular dystrophy is a very serious
muscle disorder.

cardiac muscle

fracture

joint

ligament

marrow

osteoporosis

Vocabulary Review

Complete each sentence with a term from the list.

1. A _____ is a type of tissue that holds bones together at a joint.

2. The muscle tissue that makes up the heart is called _____.

3. The soft tissue inside bones that makes blood cells is called _____.

4. A _____ is a place in the body where two or more bones meet.

5. A break or crack in a bone is called a _____.

6. A condition that causes weak and brittle bones is _____.

Chapter Quiz

Write your answers on a separate sheet of paper. Use complete sentences.

1. What are the jobs of bones and muscles?

2. What minerals are found in bone? What do these minerals do for your bones?

3. What kind of tissue holds bones together?

4. What disorder causes joints to stiffen and swell?

5. How can you keep your bones and joints healthy?

6. Give two examples of involuntary muscles.

7. What kind of tissue connects muscles to bones?

8. What kind of muscles can you control?

9. What is muscular dystrophy?

10. How do steroids affect the body?

CRITICAL THINKING

11. Why do skeletal muscles work in pairs?

12. How is nutrition related to the health of your skeletal and muscular systems?

 Online Health Project

Healthy bones and joints are an important part of your overall wellness. Find out more about disorders of bones and joints. Write a report that includes information on causes and treatments for each disorder. Use the Internet, encyclopedias, or other resources to research disorders of bones and joints.

HEALTH LINKS™

Go to www.scilinks.org/health. Enter the code **PMH200** to research **disorders of bones and joints**.

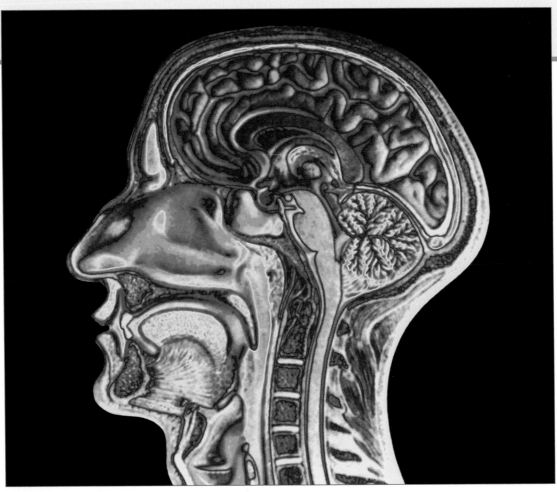

Your brain receives and interprets messages from other parts of your body.

Learning Objectives

- Describe the structure and function of the nervous system and the endocrine system.

- Describe the structure and function of the five sense organs.

- Describe disorders of the nervous system and endocrine systems.

- Name several ways to keep your nervous and endocrine systems healthy.

- LIFE SKILL: Describe how you can be an advocate for blind and deaf people in your community.

Chapter 2 ▷ The Nervous and Endocrine Systems

Words to Know

cerebrum	(suh-REE-bruhm) the part of the brain that controls voluntary muscle movements, thinking, learning, memory, speech, and the senses
cerebellum	(suhr-uh-BEHL-uhm) the part of the brain that controls balance and coordination
medulla	(mih-DUHL-uh) the part of the brain that controls involuntary functions
neuron	(NOOR-ahn) a nerve cell
dendrite	the fiber on neurons that carries messages into the cell
axon	the fiber on neurons that carries messages away from the cell
cornea	the clear, curved covering on the outer surface of the eye
iris	the colored part of the eye that controls the size of the pupil
pupil	the opening in the eye that lets light in
lens	the part of the eye that focuses light on the retina
retina	the layer of sensory neurons at the back of the eye that detects light
hormone	a type of chemical messenger in the body that is produced by endocrine glands

The Nervous System

Have you ever answered a ringing telephone, thrown a ball, or solved a difficult math problem? These activities are all controlled by your nervous system. Your nervous system allows you to think, speak, see, move, and feel.

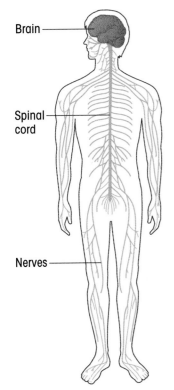

Brain

Spinal cord

Nerves

The nervous system includes the brain, spinal cord, and nerves.

Look at the drawing to the left. Your nervous system is made up of your brain, spinal cord, and a network of nerves. These organs are all made of nerve cells. The nervous system also includes your sense organs.

The Brain and Spinal Cord

The main job of the brain is to receive, interpret, and send impulses to the rest of your body. The brain is made up of more than 160 billion nerve cells.

The Parts of the Brain

There are three main parts to the brain. The largest part of the brain is the cerebrum. The **cerebrum** controls many voluntary muscle movements. It also controls thinking, learning, memory, speech, and decision making. Another job of the cerebrum is to interpret messages from the sense organs.

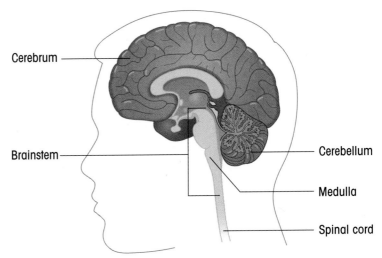

Cerebrum

Brainstem

Cerebellum

Medulla

Spinal cord

The brain has three major parts: the cerebrum, cerebellum, and brainstem.

The part of the brain that is located just below the back of the cerebrum is the cerebellum. The **cerebellum** helps to coordinate muscle movements. The cerebellum also helps you to keep your balance.

The *brainstem* is the third part of the brain. It connects the brain to the spinal cord. In the lower part of the brainstem, there is an organ called the medulla. The **medulla** controls involuntary body functions such as breathing, digestion, and heart rate. Look at the table below to review the main parts of the brain.

Parts of the Brain	
Cerebrum	• Largest part of the brain • Controls voluntary movements, thinking, and memory
Cerebellum	• Located in the back of the head • Helps with coordination and balance
Brainstem	• Connects the brain to the spinal cord • Includes the medulla, which controls involuntary body function

The human brain has three main parts.

Health & Safety Tip

Like other parts of your body, your brain can be injured in an accident. Protect your brain by wearing a helmet when performing activities in which you could hurt your head. These activities include biking, skating, skiing, snowboarding and football.

The Spinal Cord

The spinal cord is a ropelike structure that is made up of many nerve cells. This cord runs from the base of the brain down to the lower part of your back. It is protected by your backbone.

The spinal cord acts as the nervous system's center for processing information. Most of the messages between your brain and the rest of your body must travel along the spinal cord. There are 31 pairs of nerves that extend from your spinal cord to all parts of your body.

✓ Check Your Understanding

Write your answers in complete sentences. Use a separate sheet of paper.

1. What are the parts of the nervous system?

2. What is the function of the spinal cord?

3. What are the three main parts of the brain?

4. CRITICAL THINKING Which part of the brain are you using when you memorize a number?

People in Health

ALANA SHEPHERD—ADVOCATE FOR PEOPLE WITH A SPINAL CORD INJURY

Alana Shepherd is not a doctor or a famous person, but she has made a difference to many people with a spinal cord injury.

In 1973, Alana's son, James, injured his spinal cord in a surfing accident. At the time, there were no hospitals that could treat his type of injuries in the southeastern United States, where they lived. Because he could not get the care he needed at home, James had to travel a great distance to recover. Alana set out to make sure other families in the region would find the care they needed closer to home.

Two years later, Alana and her husband, Harold, started the Shepherd Center in Atlanta, Georgia. The Shepherd Center is a hospital that treats people with a spinal cord injury or other nervous system disorders. Its mission is to help people with spinal cord injuries become as independent as possible. Thanks to Alana Shepherd's hard work, the Shepherd Center is the largest hospital in the United States that treats spinal cord injuries.

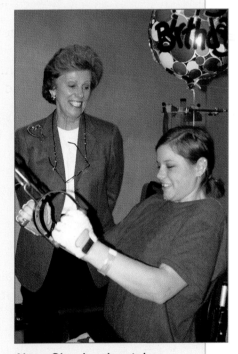

Alana Shepherd watches a patient use an exercise machine.

CRITICAL THINKING How did Alana Shepherd's experience make her want to help others?

Nerve Cells

A **neuron** is a nerve cell. Neurons come in many sizes and shapes. In fact, one neuron in your leg can be more than 3 feet long! Neurons are the basic units of structure that make up the nervous system.

A typical neuron has three main parts. The largest part of the neuron is the cell body. There are long fibers that branch from the cell body. Some of these fibers carry messages from other neurons into the cell body. These fibers are called **dendrites**. Other fibers carry messages away from the cell body to other neurons. These fibers are called **axons**. Look at the drawing to the right to see the main parts of a neuron.

Sensory neurons are connected to your sense organs. The eyes, ears, nose, tongue, and skin are your sense organs. *Sensory neurons* take messages away from the sense organs to the brain and the spinal cord.

Motor neurons carry messages away from the central nervous system to muscles and glands. *Glands* are organs that make chemical substances that are used or released by the body. Muscles and glands respond to information sent from the brain.

A neuron is a nerve cell.

Synapses

When an impulse travels along a neuron, the axon of one neuron usually does not actually touch the dendrite of the next neuron. The spaces between neurons are called *synapses*. Impulses must "jump" across this space. Chemicals released by the axon carry impulses across the synapse to the dendrite of the next neuron. Then, the impulse continues traveling along the next neuron.

The Nervous System and Your Actions

The nervous system sends messages among the different organs of the body. These messages are signals called *impulses*. There are three ways the nervous system processes impulses.

Reflexes are involuntary responses to outside stimuli. They happen without your brain "thinking" about them. These impulses are processed in your spinal cord rather than in your brain.

Suppose you touch a very hot pan. Your arm muscles lift your hand away from the pan. This movement occurs before the pain message reaches your brain.

Talk About It

Reflexes help a person survive. Explain why this is true.

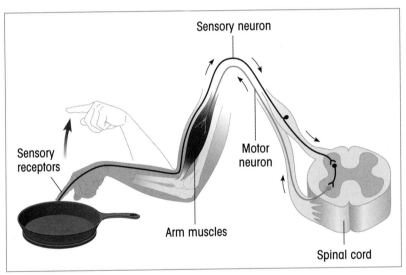

Sensory neuron

Sensory receptors

Motor neuron

Arm muscles

Spinal cord

Reflexes cause your muscles to pull your finger away from a hot pan involuntarily.

Automatic impulses are those that control how your body's organs function. They cause your stomach muscles to squeeze together. They help you breathe. They also keep your heart beating.

Voluntary nerve action means that you think about doing something first. Then, you do it. Voluntary nerve impulses usually control muscles. The steps below show the nerve impulses that take place when you hear a car horn. Even though these nerve impulses involve many different organs and types of cells, they happen very fast.

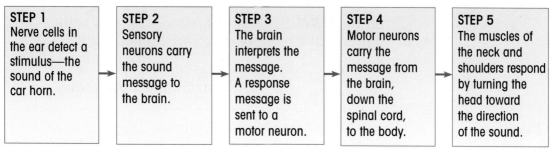

STEP 1	STEP 2	STEP 3	STEP 4	STEP 5
Nerve cells in the ear detect a stimulus—the sound of the car horn.	Sensory neurons carry the sound message to the brain.	The brain interprets the message. A response message is sent to a motor neuron.	Motor neurons carry the message from the brain, down the spinal cord, to the body.	The muscles of the neck and shoulders respond by turning the head toward the direction of the sound.

Impulses travel in a series of steps.

Disorders of the Nervous System

A stroke happens when something prevents the brain from getting its blood supply. Without blood, nerve cells in the brain die. Many people become partly paralyzed from having a stroke.

Cerebral palsy is another disorder caused by damage to the brain. A person with cerebral palsy has less-than-normal control over his or her muscles. This person's hands may shake, or he or she may not be able to form words. In some cases, the person's arms are stiff and slow-moving. Cerebral palsy is sometimes caused by poor prenatal care. *Prenatal* means before birth, so prenatal care is care received by a baby while it is still inside its mother.

A third disease that affects the nervous system is *multiple sclerosis*. This disease attacks the body's nerves. The damaged nerves are then unable to send or receive messages. Multiple sclerosis usually attacks people between the ages of 20 and 40.

Health Fact

A headache is a disorder of the nervous system. Headaches may be caused by stress, tension, food allergies, or physical illness. Migraine headaches are more serious. People with migraines sometimes get upset stomachs, and may be bothered by light and sound. Migraines can last for several days.

Epilepsy is a disorder that causes a person to have seizures. People suffering from this condition have abnormal brain waves.

Drugs and alcohol also affect the nervous system. A person's eyesight, thinking, and muscle control are impaired after drinking alcohol or abusing drugs. Drugs and alcohol often have long-term effects, too. Memory loss, nervousness, and poor emotional health are some of these effects. There may also be signs of brain damage as a result of drug and alcohol abuse.

Keeping Your Nervous System Healthy

Here are some tips you can use to keep your nervous system healthy.

- Be careful. Head and back injuries can cause lasting damage to the nervous system. Use your common sense to prevent accidents. Drive safely and always wear your seat belt. If you ride a motorcycle or a bicycle, wear a helmet.

- Be aware of what drugs and alcohol can do to your nervous system. Every time a person drinks, brain cells are killed. Every time a drug is abused, the chemistry of the brain is changed. Long-lasting effects to the nervous system may be serious and even deadly.

- If you become a parent, take care of your baby's nervous system by getting good prenatal care. What a pregnant woman does, even in the earliest days of pregnancy, can affect her baby's nervous system. Make sure you know the effects of diet, exercise, drug use, drinking, and smoking on your baby.

- Relax and enjoy life. Stress and tension tire and weaken your nervous system. Stress can also cause headaches. Take at least 20 minutes out of the day to exercise, nap, or just relax.

The Five Senses

People enjoy tasting food, listening to music, touching a puppy's fur, smelling flowers, and watching movies. These activities are possible because of five important organs: tongue, nose, skin, eyes, and ears.

Each sense organ has special parts and cells that allow it to work. These organs do more than help you enjoy the world. They send very important information to your brain every second. This information helps you to stay healthy.

Tongue and Taste

Your tongue is covered with sensory neurons that sense taste. These neurons are contained in taste buds. You have four kinds of taste buds—sour, sweet, salty, and bitter. The "tongue map" below shows where each of these kinds of taste buds is often found.

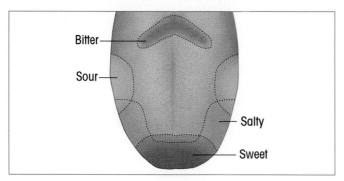

The tongue senses four types of tastes. Different parts of the tongue are more responsive to certain tastes.

Nose and Smell

Much of your sense of taste is really smell. As you eat, scents from your food enter your nose. Smells are picked up by sensory neurons in the nose. These cells have long, hairlike structures. The hairlike structures are connected to nerves. The nerves carry the smell message to the brain. The brain interprets the smell.

Health Fact

When you have a head cold, most foods have little flavor. That is because your nose is clogged. You are not smelling the food.

Write About It

Eyelids, eyebrows, eyelashes, and tears all protect your eyes. Tears come from a gland above each eye. What kinds of things do you think tears protect the eyes from?

Eyes and Sight

Light enters the eye through a clear, curved covering called the **cornea**. Behind the cornea is a colored part called the iris. The **iris** surrounds an opening in the eye called the pupil. The **pupil** allows light to enter the eye.

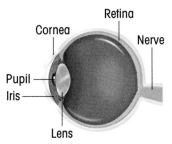

The eye is the organ of sight.

Once light enters your eye, it passes through the **lens**. The lens is clear. It bends the light rays so that they correctly hit the retina. The **retina** is the layer of sensory neurons at the back of your eye. It can sense brightness and color.

The retina is attached to a thick bundle of nerves. The nerves carry the information to your brain. Your brain then makes sense of the information and you see a picture.

Ears and Hearing

The ears are the organs of hearing. They detect sound. The ear is divided into three sections: the outer ear, the middle ear, and the inner ear.

Sound waves enter the outer ear and travel through the ear canal. The waves pass into the middle ear through the eardrum. The *eardrum* is a thin layer of tissue. When sound hits the eardrum, it vibrates. This vibration travels through small bones in the inner ear. Then, sound travels to the brain through nerves.

The ear is the organ of hearing.

Your ears also help to maintain your balance. There are three looped tubes in your inner ear. They are filled with liquid. As the liquid moves, receptor cells send information about balance to your brain.

Skin and Touch

Skin is the sense organ of touch. It can sense heat, cold, pressure, and pain. Some parts of your skin can sense touch better than others. Your fingertips have many more sensory neurons than your arms do.

Disorders of the Sense Organs

Problems with sense organs include poor vision, cataracts, ear infections, and deafness.

Nearsighted and Farsighted Eyes

Most eye problems occur when the lens does not focus the "picture" exactly on the retina. A nearsighted person can see close objects clearly. Objects that are far away look fuzzy and blurred. Because the eyeballs are too long, a picture is focused in front of the retina, instead of on the retina.

To a farsighted person, close objects look fuzzy, and faraway objects look clear and crisp. The eyeballs of the farsighted person are too short. A picture is focused behind the retina, instead of on the retina.

Eyeglasses can correct both these problems by adding an extra lens to each eye. The extra lenses move a picture backward or forward so that it is focused exactly on the retina. Look at the drawings at the right.

Contact lenses are small lenses that are worn on the cornea. They fit right over the pupil. Contact lenses may cost more than most glasses. Also, they are not as easy to take on and off as glasses. But contact lenses provide the wearer with better overall vision, and they cannot be seen.

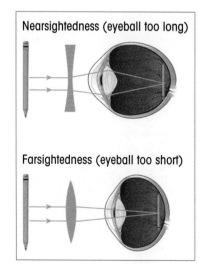

Nearsightedness (eyeball too long)

Farsightedness (eyeball too short)

Eyeglass lenses help people who are nearsighted and farsighted to see better.

More Serious Eye Disorders

A disorder that is common among older people is *cataracts*. This is a condition that causes the lens of the eye to become blurry. Cataracts can cause blindness. Usually, surgery can repair this problem.

As you get older, eye exams become more and more important. Certain eye problems happen more often in older people. One such problem is called glaucoma. *Glaucoma* causes painful pressure in the eyeball and can cause loss of sight. Glaucoma can be treated with medicines if it is caught in time.

Some Common Ear Problems

One of the most common ear disorders is an ear infection. Ear infections are caused by bacteria or viruses. Most ear infections are not serious. They can be treated with medicine.

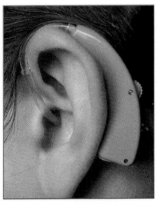

Hearing aids, like the one shown here, can help people with a hearing loss hear better.

Other hearing problems are harder to cure. People can lose some or all of their hearing after listening to music that is too loud or working with loud machinery. Sometimes, babies are born with damaged ear parts. This damage can cause loss of hearing or even total deafness. Operations can help and so can hearing aids. One kind of hearing aid sends sounds directly to the inner ear. Another kind sends the sound waves through the skull bones just behind the ear. Then, the inner ear picks up the sound.

Some problems with the ear can disrupt a person's balance. For example, *vertigo* is the feeling of dizziness. It can occur when a person rises too quickly from a sitting position, which affects the fluid in the inner ear.

Keeping Your Sense Organs Healthy

Here are a few tips for keeping sense organs healthy:

- Have ear and eye exams every year.
- Watch television from a safe distance, usually a distance of five times the width of the screen.
- Wear goggles when using sharp objects and chemicals.
- Use eyewash if something gets into your eye.
- Do not listen to music that is too loud.
- Do not place cotton swabs deep into your ear.
- See a doctor if you have an earache.

Talk About It

Motion sickness can be caused by a disturbance in the inner ear. Some people get motion sickness when driving in a car, riding a bus, or being on a boat. Do you ever get motion sickness? What symptoms have you had?

The Endocrine System

There are many different endocrine glands scattered throughout the human body. Together, they make up the endocrine system. The job of the endocrine system is to help control body functions. Read the organs in the drawing on the right.

Hormones: Another Kind of Messenger

Some body functions and messages are controlled by nerve impulses. Others are controlled by chemical messengers. Endocrine glands produce chemical substances that are released directly into the bloodstream. These chemical substances are called **hormones**. Hormones regulate body functions.

Different endocrine glands produce different hormones. Each gland has different functions. For example, the thyroid releases a hormone that controls how quickly the body uses energy. The pancreas makes a hormone that controls the amount of sugar in the blood. Endocrine glands release their hormones into the space that surrounds them. Then, the hormones pass into blood vessels. The blood carries the hormones throughout the body to the correct cells.

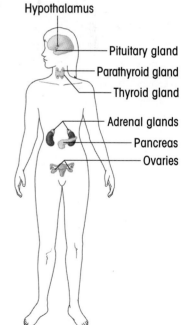

Hypothalamus

Pituitary gland

Parathyroid gland

Thyroid gland

Adrenal glands

Pancreas

Ovaries

The female endocrine system

Hormones at Work

The endocrine system plays an important part in maintaining a healthy body. Hormones produced by endocrine glands help to keep the body's cells and tissues in balance. Endocrine glands accomplish this balance by making sets of hormones that have opposite effects. In this way, the two hormones balance each other's effects.

Disorders of the Endocrine System

Most disorders of the endocrine system are caused by one or more of the glands not working correctly. A gland may produce too much, or not enough of a hormone. This imbalance can have harmful effects on the rest of the body.

The pituitary gland produces important hormones that control other glands and hormones. If the pituitary gland does not work properly, other parts of the body are affected.

For example, the pituitary gland may make too little thyroid-stimulating hormone, or TSH. When too little TSH is made, the thyroid gland does not make enough of its own hormones. The rate that the body uses energy is then lowered. A person can then gain weight and have a lower heart rate and body temperature.

Other endocrine problems include infertility, growth and development disorders, diabetes, hypoglycemia, and kidney disorders. The part of the body that is affected depends on which endocrine gland has a disorder.

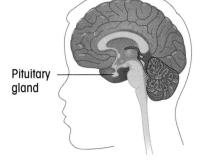

Pituitary
gland

A pituitary gland that does not work correctly may cause a disorder of the endocrine system.

✓ **Check Your Understanding**

On a separate sheet of paper, write a paragraph that explains how the endocrine system helps to maintain a balance in the body.

LIFE SKILL
Advocating for Blind and Deaf People in Your Community

Have you ever seen a blind or deaf person walking with a service dog? Service dogs help people who are blind or deaf to lead more independent lives. Not all dogs can be service dogs. They must receive special training. The training begins when the dogs are puppies.

Groups that provide service dogs rely on many volunteers. They need people to raise the young puppies in their homes. The pups stay until they are about 18 months old. Raising a service dog is a big commitment. It requires a lot of time and loving care. Sometimes it is hard to see the dog leave, but it must. The dog's next step is a special training school to learn how to help a blind or deaf person. The blind or deaf person also must learn how to work with the dog.

Many groups that provide service dogs also need financial help. People who use service dogs do not have to pay very much for them. Donations support many of the training programs.

You can help the blind and deaf in your community by supporting service dog groups. Read the examples listed below. Which ones interest you?

Guide dogs get special training to help blind people.

- Raising a service dog puppy in your home.
- Recruiting other service dog puppy raisers.
- Raising funds for service dog training programs.

Design a brochure that encourages support for a service dog group. Be sure to explain why people should support it. Also explain different ways people can support the group.

Applying the Skill
Interview someone who works at a service dog group. Find out what types of donations or volunteer work they need. Then, organize your classmates and help out.

Summary

The nervous system is made up of the brain, spinal cord, and nerves.
Disorders of the nervous system include strokes, cerebral palsy, epilepsy, and multiple sclerosis.
You can keep your nervous system healthy by preventing accidents, avoiding drugs and alcohol, eating healthy, and getting enough sleep.
The five senses include taste, smell, touch, sight, and hearing.
Some disorders of the senses include poor vision, cataracts, blindness, loss of hearing, and deafness.
You can keep your sense organs healthy by getting regular exams, not watching television too closely, and not listening to music that is too loud.
The endocrine system includes endocrine glands and hormones. Disorders of the endocrine glands can have harmful effects on many other parts of the body.

cerebrum

medulla

hormone

lens

neuron

Vocabulary Quiz
Complete each sentence with a term from the list.

1. The part of the brain that controls voluntary muscle movements, thinking, learning, memory, and the senses is the _____.

2. The part of the brain that controls involuntary functions is the _____.

3. A nerve cell that carries impulses to and from the brain is called a _____.

4. The part of the eye that focuses light on the retina is the _____.

5. A chemical messenger that is released by endocrine glands into the bloodstream is a

_____.

Chapter Quiz

Write your answers on a separate sheet of paper.
Use complete sentences.

1. What is the function of a neuron?

2. What are the functions of the three main parts of the brain?

3. What is the function of the spinal cord?

4. What are five senses and their organs?

5. Which organ interprets most of the information taken in by the senses?

6. What does light pass through when it enters the eye?

7. What are some ways you can keep the sense organs healthy?

8. Name three endocrine glands.

CRITICAL THINKING

9. How are nerve impulses and hormones similar? How are they different?

10. Why do you think it is important to protect your brain and spinal cord by using safety equipment such as seat belts and helmets?

Online Health Project

Diabetes is a disease that affects the lives of many children and young adults. New research studies are being done to find better treatments or even a cure for diabetes. Find out more about current diabetes research. List and explain the treatments or possible cures in a brochure or poster.

HEALTH LINKS.
Go to www.scilinks.org/health.
Enter the code **PMH210** to research **diabetes**.

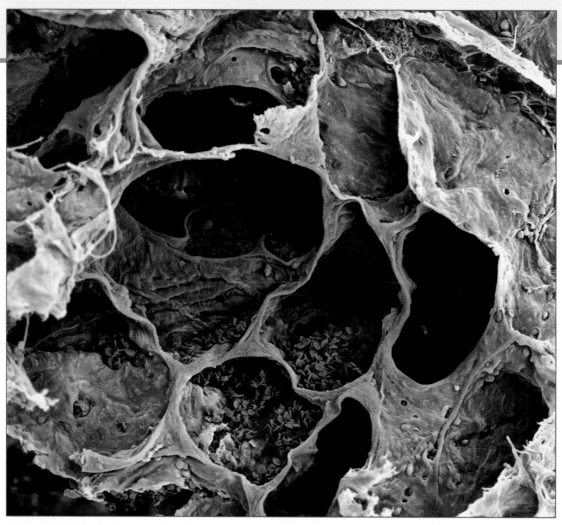

This photograph shows red blood cells located inside the air sacs of the lungs.

Learning Objectives

- Describe the structure and function of the circulatory, lymphatic, immune, and respiratory systems.

- Describe disorders of the circulatory, lymphatic, immune, and respiratory systems.

- Explain how you can keep your circulatory, lymphatic, immune, and respiratory systems healthy.

- LIFE SKILL: Analyze how culture can influence a person's health.

Chapter 3

The Circulatory, Immune, and Respiratory Systems

Words to Know

circulatory system	the organ system that moves blood throughout the body
vein	a blood vessel that carries blood toward the heart
artery	a blood vessel that carries blood away from the heart
capillary	a tiny blood vessel that connects arteries to veins
plasma	the liquid part of blood that carries nutrients and wastes throughout the circulatory system
platelet	a piece of a cell found in blood that helps blood cells clump together
red blood cell	a type of cell found in blood that carries oxygen to cells
white blood cell	a type of cell that helps the body fight infection
hypertension	high blood pressure
cholesterol	a fatlike substance needed by the body in small amounts
antibody	a molecule that attaches to a specific pathogen
immunity	the ability to resist a certain disease
vaccine	a substance that stimulates immunity to a disease

The Needs of Cells

All cells need nutrients and oxygen. These materials help cells to carry out life functions. Cells also produce wastes. Cells need to get rid of these wastes.

In this chapter, you will read about organ systems that help your cells get the materials they need and how they help your body fight disease.

The Circulatory System

Your cells need certain materials. These materials are brought to your cells by blood. Blood flows throughout your body. It carries oxygen and nutrients to all your body cells.

When cells carry out life functions, they also produce wastes. Some of these wastes include carbon dioxide and water. Blood carries these wastes away from body cells.

The **circulatory system** moves blood throughout the body. This organ system includes the heart, blood vessels, and blood. Look at the drawing on the left.

Arteries

Veins

Heart

The circulatory system is made up of the heart, blood vessels, and blood.

The Heart

The heart is the main organ of the circulatory system. It is about the size of a closed fist. The heart is made up of cardiac muscle. The heart works by pumping and relaxing. These actions keep blood flowing throughout the body. The heart is divided into four parts, or *chambers*. The two top chambers are called *atria* (singular *atrium*). The two bottom chambers are called *ventricles*. Valves open and close to control the direction of blood flow.

The drawing on page 41 shows the path of blood through the heart. First, blood flows into the right atrium. From there, it is pumped into the right ventricle. Then, arteries carry the blood to the lungs. In the lungs, the blood picks up oxygen. Now, the oxygen-rich blood returns to the left atrium through large veins. The heart pumps the blood into the left ventricle. From there, it moves through the aorta and out to the body. The blood delivers nutrients and oxygen to all cells.

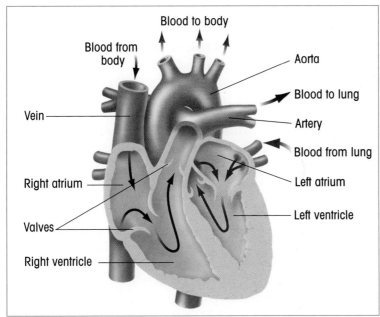

Blood to body

Blood from body

Aorta

Blood to lung

Artery

Vein

Blood from lung

Right atrium

Left atrium

Left ventricle

Valves

Right ventricle

The human heart has four chambers and four valves.

Write About It

Many people have surgery to fix damaged heart valves. Why are the valves so important to the health of the heart? Write your answer on a separate sheet of paper.

Blood Vessels

Veins are blood vessels that carry blood toward the heart. **Arteries** are blood vessels that carry blood away from the heart. **Capillaries** are tiny blood vessels that connect arteries to veins.

Oxygen and nutrient molecules pass through capillary membranes into body cells. At the same time, waste molecules from the body cells pass into the blood. These wastes are carried through veins back to the heart into the lungs. You exhale some of the wastes into the air.

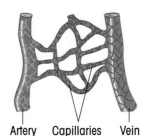

Artery   Capillaries   Vein

Capillaries connect arteries to veins.

✓ Check Your Understanding

Write your answers in complete sentences.

1. What does the circulatory system do?

2. What parts make up the circulatory system?

3. Name three types of blood vessels.

Blood

Blood transports nutrients, oxygen, and wastes, and fights infection. About 55 percent of blood is a liquid called **plasma**. This liquid carries nutrients and wastes throughout the circulatory system.

There are three kinds of solids in blood: platelets, red blood cells, and white blood cells. **Platelets** are actually pieces of cells. They help blood cells to clump together. This action helps wounds to stop bleeding.

Red Blood Cells

A protein in **red blood cells** delivers oxygen to other body cells. This protein, called *hemoglobin*, contains the mineral iron. Iron gives blood cells their red color. The more oxygen the protein carries, the brighter the color of the cells.

White Blood Cells

White blood cells fight infection. They destroy harmful bacteria and defend the body against other harmful substances. You have fewer white blood cells than red blood cells. There is about one white blood cell for every 600 red blood cells. But, when your body is fighting an infection, the number of white blood cells in your body can quickly increase.

Blood Types

There are four main blood types: A, B, AB, and O. These blood types are based on whether certain substances are present on the red blood cells. People who receive blood from a blood donor must get the right type of blood. The chart at the left shows the types of blood different people can receive from a donor.

Blood Type	Can Get Blood from People With
A	O, A
B	O, B
AB	All blood types
O	O only

Rh Factor

There are other differences among blood cells besides blood types. Sometimes there is a substance on the surface of a blood cell called *Rh factor*. Most people in the United States are called *Rh positive*. They have the Rh substance on their blood cells. People that do not have the substance on their blood cells are called *Rh negative*.

It is very important that a person getting a blood transfusion receives blood matching with their Rh. It is also important for a pregnant mother to know whether she is Rh positive or Rh negative. A difference between her blood and the unborn baby's blood can cause problems during delivery.

Talk About It

Do you know what your blood type is? Why do you think knowing your blood type is important?

Think About It

WHY IS BLOOD DONATION CRITICAL TO PUBLIC HEALTH?

According to the American Red Cross, someone in America needs blood every two seconds. The blood is used to treat accident victims and cancer patients. It is also used during surgeries. Nearly all of this blood comes from volunteer donors.

To be a blood donor, you must be in good health and at least 17 years old. While millions of Americans fit these guidelines, only about 5 percent actually donate blood. As a result, the blood supply varies from day to day. In the summer months and during holidays, fewer volunteers donate blood. But, the amount of people who need blood often increases at those times. In addition, as the population ages, the need for blood will increase. The only way to address this need is to encourage more people to save a life by becoming a blood donor.

YOU DECIDE Why is blood donation so important to public health?

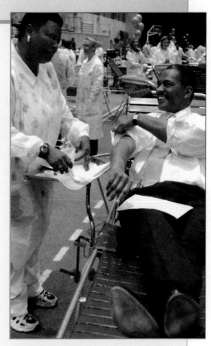

Blood drives are organized to encourage more people to donate blood.

Disorders of the Circulatory System

There are many disorders that can happen to the circulatory system. In fact, heart disease is the most common cause of death of Americans. In 2001, more than 700,000 Americans died of heart disease. The most common disorders are high blood pressure, high cholesterol, heart attacks, and strokes.

High Blood Pressure

If you have ever had a complete check-up, you probably had your blood pressure taken. A blood pressure reading measures how hard the heart is working. This reading can determine if you have high blood pressure.

High blood pressure, also called **hypertension**, is a sign that the blood vessels or heart are not working properly. High blood pressure may be caused by a variety of factors including diet, age, weight, and whether a person smokes tobacco. It can also be inherited from a previous generation. People with diabetes and kidney disease and who are pregnant are more likely to develop high blood pressure.

High Cholesterol

Blood vessels can also become narrow when cholesterol builds up on their walls. **Cholesterol** is a fatlike substance that plays an important part in the body. It can be found in foods such as butter, eggs, liver, cheese, and beef. Cholesterol helps build cell membranes and sex hormones. It also helps you to digest your food. But, too much cholesterol can build up on artery walls, blocking blood flow. Compare the photos shown at the right. This blockage can result in high blood pressure.

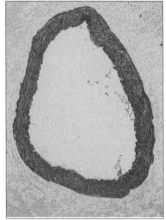
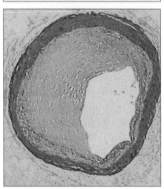

The artery on the top is healthy. The yellow part in the center is the opening that blood flows through. The artery on the bottom is blocked with cholesterol and other fats, shown in orange.

Heart Attacks

A heart attack usually begins as a pain in the chest. The pain spreads through the arms, throat, and back. The victim starts to sweat and has trouble breathing.

Heart attacks are often caused by blocked blood vessels called *coronary arteries*. When blood does not get to the heart, the heart shuts down. The heart is a muscle and is made up of living cells. It needs a supply of nutrients and oxygen in order to function. When that supply is cut off, the heart cannot keep working.

Strokes

A stroke can happen when the blood supply to part of the brain is cut off. The brain is made up of cells, just like the heart. If some of the brain cells do not get the oxygen and nutrients they need, they can stop working.

After a stroke, some people have brain damage. This damage can affect the way they talk and move. It can also affect their memory.

Other Circulatory System Disorders

Some other disorders of the circulatory system are anemia, varicose veins, and heart defects. Anemia occurs when there are low levels of red blood cells in the body. Without the appropriate number of red blood cells, the body's organs and tissues do not get enough oxygen and cannot work properly.

Varicose veins occur when a vein becomes swollen. This happens because the valves that move blood to the heart do not close properly and start to leak. A heart defect is an abnormality of the heart's structure or function. Heart defects are disorders people are born with. Approximately 30,000 to 40,000 (or 1 in 1,000) children are born each year with a heart defect. Some types of heart defects can be corrected by surgery.

Talk About It

Blood vessel diseases, such as blocked arteries, kill more Americans than all other causes of death combined. More than 30 percent of Americans over the age of 50 have high blood pressure. What do you think this says about the diets of these Americans?

Keeping Your Circulatory System Healthy

Here are the best ways to keep your heart strong and your blood vessels clear and flexible:

- Get plenty of exercise. You should get at least 15 to 30 minutes of exercise every day.

- Eat foods that are low in fat and cholesterol. Fish, skinless chicken, tofu, beans, fruits, and vegetables are good for the blood and heart. Butter, eggs, cheese, and animal fats, including meat, are high in cholesterol.

- Learn how to read food labels. Foods that contain "unsaturated fats" are healthier for you than foods that contain "saturated fats."

- Put less salt on your food. Salt can increase your blood pressure.

- Stay away from tobacco. Smoking leads to high blood pressure and a weak heart. Smoking also increases your risk of lung cancer.

- Get your cholesterol and blood pressure checked regularly.

- Ask a doctor about the normal blood pressure for a person in your age group.

✓ Check Your Understanding

Write your answers in complete sentences. Use a separate sheet of paper.

1. What are three common disorders of the circulatory system?

2. What substance can cause blocked blood vessels?

3. What factors are related to high blood pressure?

4. **CRITICAL THINKING** How do your actions affect the health of your circulatory system?

The Lymphatic System

As blood moves through the body, some fluid leaks into body tissues. As much as three quarts of fluid can leak into body tissues per day.

The lymphatic system collects this fluid and returns it to the circulatory system. The clear fluid, called *lymph*, contains special white blood cells. These white blood cells, called *lymphocytes*, fight disease. Lymph moves through vessels and special organs, including the tonsils, thymus, spleen, and lymph nodes shown in the drawing below.

 Health Fact

The lymphatic system does not have a pump like the heart. Pressure from blood and the squeezing of skeletal muscles move lymph through lymph vessels.

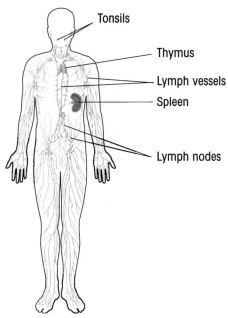

Tonsils

Thymus

Lymph vessels

Spleen

Lymph nodes

The lymphatic system returns fluid back to the circulatory system.

Lymph is filtered by the lymph nodes and tonsils. When you have an infection, your lymph nodes or tonsils can become larger. They become filled with a large number of white blood cells that are fighting the bacteria or virus causing your infection. The thymus gland helps certain white blood cells to grow. The spleen destroys damaged red blood cells.

Talk About It

Have you ever heard a doctor use the term swollen glands? What do you think the doctor is referring to?

The Immune System

The immune system defends against pathogens, or disease-causing substances. The immune system contains different kinds of cells and molecules that fight against pathogens in different ways.

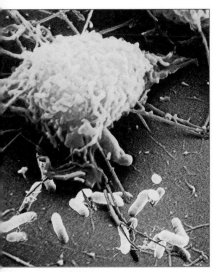

Write About It

How do you think pathogens get into the body?

White Blood Cells

One type of cell that is part of the immune system is the white blood cell. For example, when a pathogen enters the body, the white blood cells find it. They enter the tissue that is infected and attack the pathogen.

There are two special kinds of white blood cells (T cells and B cells) called lymphocytes. They are part of the lymphatic system. They are also part of the immune system because they help the body fight disease. T cells identify pathogens. Then, they may attack the pathogens. T cells also attack the damaged body cells that the pathogens have already harmed. Other times, T cells do not attack the pathogens directly. Instead, they stimulate B cells to produce substances called antibodies.

Antibodies find certain pathogens and attach to them like a flag. This attachment makes it easier for other cells of the immune system, such as white blood cells, to identify the pathogens. Then, the other cells find the pathogens and attack them.

The white blood cell, shown in blue, is attacking E. coli bacteria, shown in yellow.

Vaccines and Immunity

The ability to resist a certain disease is called **immunity**. People are often born with certain immunities. These are called natural immunities. Acquired immunities are immunities that people develop or acquire over time.

One type of acquired immunity can occur when a mother passes antibodies for certain diseases to her developing baby. Another form of acquired immunity happens when you are infected with a disease only once. Then, you will not get the disease again. Your body produces antibodies to the disease.

Getting a vaccine is another form of acquired immunity. A **vaccine** is made from dead or weakened forms of pathogens. After receiving a vaccine, the body develops antibodies for the disease. This vaccination makes the body immune to that disease.

This child is getting a vaccine.

Health and Technology

VACCINE PRODUCTION

Vaccines are produced in several different ways. One way is by using a dead or weakened form of a virus that causes a disease. When this form of the virus is injected into a person, it causes the person to become immune to the disease. If that person is later infected with the virus, they will not get sick. The flu vaccine is made in this way.

Genetic engineering is also used to make vaccines. *Genetic engineering* is the changing of an organism's genetic material, or genes. To produce a vaccine through genetic engineering, researchers remove a gene or genes from a virus. The genetic material is placed into a host cell, which can be a bacteria or yeast.

Vaccines can be genetically engineered.

The host cell grows rapidly and makes large amounts of the proteins from the virus that causes disease. Then, a vaccine is made with these proteins. This method has been used successfully in the production of a vaccine for hepatitis B.

CRITICAL THINKING Why do organisms produced for vaccines contain proteins that cause the immune system to respond?

The Respiratory System

The respiratory system moves oxygen into the body and carbon dioxide out of the body. This system includes the lungs and the pathways that air moves through. The drawing on page 51 shows the parts of the respiratory system.

Breathing Air In and Out

The air you breathe contains oxygen. It can go into your nose. The small hairs and mucus in your nose filter out dust and dirt. You can also breathe in air through your mouth. But your mouth cannot clean the air as well as your nose can. As you breathe out, air rushes past your vocal cords.

The Diaphragm

As you breathe in, the muscles in your chest lift your rib cage up and outward. At the same time, a large skeletal muscle, called the *diaphragm* (DY-uh-fram), pulls the bottom part of your chest cavity downward. Both actions increase the amount of space in your chest cavity so that air can move into your lungs. When you breathe out, the rib muscles and diaphragm relax. The size of your chest gets smaller, and air is forced out.

Tubes and Pathways

Inhaled air moves into your throat. Both air and food pass down the throat. Food continues through it to another tube that leads to the stomach.

Next, air moves into the trachea. The *trachea* (TRAY-kee-uh), or windpipe, brings air from your throat into your lungs. When you swallow, a flap of tissue called the *epiglottis* (ehp-uh-GLAHT-ihs), closes. It keeps food and liquids out of the trachea.

The trachea divides into two smaller tubes. These smaller tubes are called *bronchi* (BRAHN-ky). Each of these tubes (a bronchus) goes into a lung. Air moves through these small tubes and into your lungs.

Health Fact

Sneezing is the respiratory system's way of getting rid of germs. "Sneezed" air can travel more then 100 miles per hour.

Lungs

You have two lungs. They are big, spongy organs. In each lung, the bronchus branches into smaller and smaller tubes. These tubes end in millions of tiny air sacs. These air sacs are called *alveoli* (al-VEE-uh-ly).

In these sacs, gases are exchanged. Oxygen passes into capillaries. The waste gas carbon dioxide passes from the blood into the air sac. You get rid of carbon dioxide when you breathe out, or exhale.

 **Health Fact**

Each human lung has about 150 million air sacs. All together these air sacs have a surface area that is about 40 times that of your skin.

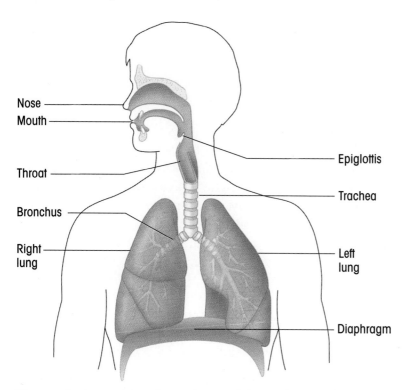

Nose
Mouth
Throat
Bronchus
Right lung
Epiglottis
Trachea
Left lung
Diaphragm

The respiratory system is made up of the pathways that air moves through and the lungs.

Disorders of the Respiratory System

Colds and flus are the most common disorders of the respiratory system. These disorders are infections that are caused by viruses and bacteria. They irritate your lungs and cause a buildup of mucus. Severe colds and flus can cause pneumonia, a condition that fills the lungs with fluid. Medicines can help cure this condition. If untreated, pneumonia can be deadly. Allergies, such as hay fever and asthma, are also common respiratory disorders.

More serious disorders of the respiratory system include *emphysema* and lung cancer. These diseases are both very serious. They are often caused by smoking. The tar found in cigarettes can clog the air sacs in the lungs, making it harder to breathe. This damage to the lungs can then lead to lung cancer.

Keeping Your Respiratory System Healthy

Here is a list of tips you can use to keep your respiratory system healthy:

- Do not smoke.
- If you get sick with a cold or flu, get plenty of rest. Drink a lot of liquids. If your symptoms do not go away, see a doctor.
- Get a check-up once a year even if you feel healthy.

✓ Check Your Understanding

Answer the questions in complete sentences.

1. What are the parts of the respiratory system?

2. How does smoking affect the lungs?

LIFE SKILL
Analyzing Influences of Culture on Health

Culture and health may not seem related, but often they are. People in one culture may be at greater risk for certain diseases. Genes, diet, and access to healthcare can play a role in your health. Your culture may also affect what you do when you get sick.

In some cultures, health matters are private. People may not complain about being sick, and they might be afraid to seek help. Some people go to traditional healers for help. A patient may see a healer for some problems and a doctor for others.

Understanding cultures can be helpful when it comes to health. If your culture has a high rate of a certain disease, you can take steps to reduce your risk. You also can think about new ways to treat a problem.

HOW CULTURE AFFECTS HEALTH

- African American people have a higher rate of death from heart disease.
- Americans have more overweight people than most other cultures.
- Rural Chinese cultures have very few cases of osteoporosis, a bone disease.

People in a culture have many things in common. But not all people fit the pattern of their culture. Each person has unique health issues and treatment ideas.

Answer the following questions using complete sentences.

1. How can knowing about your culture help you with your health?

2. How can a culture affect what a person does when he or she is sick?

Applying the Skill

Think about your family and your culture. Write about a cultural factor that affects your health.

Summary

The circulatory system moves blood throughout the body. It is made up of the heart, blood vessels, and blood. Disorders of the circulatory system include high blood pressure and heart disease.

The lymphatic system includes lymph, lymph vessels, and lymph nodes. Lymph contains white blood cells that fight disease.

The immune system protects the body from disease. It is made up of white blood cells and special molecules called antibodies.

The respiratory system delivers oxygen to the blood. It is made up of the lungs and pathways that air moves through. Disorders of the respiratory system include colds, asthma, emphysema, and cancer.

You can keep your body system healthy by eating a healthy diet, exercising, and not smoking tobacco.

circulatory system

immunity

plasma

vein

vaccine

Vocabulary Review

Complete each sentence with a term from the list.

1. The liquid part of blood that carries nutrients and wastes throughout the circulatory system is called _____.

2. The organ system that moves blood throughout the body is called the _____.

3. A blood vessel that carries blood toward the heart is called a _____.

4. The ability to resist a certain disease is called _____.

5. A substance that stimulates immunity to a disease is called a _____.

Chapter Quiz

Write your answers on a separate sheet of paper. Use complete sentences.

1. What are the parts that make up the circulatory system?

2. What are the two top chambers of the heart called? What are the two bottom chambers of the heart called?

3. What are some disorders of the circulatory system?

4. What are some risk factors for heart disease?

5. What are some parts that make up the lymphatic system?

6. How does the immune system fight disease?

7. What parts make up the respiratory system?

8. What is another name for the windpipe?

9. What are some common disorders of the respiratory system?

10. Why is smoking harmful to the respiratory system?

CRITICAL THINKING

11. How does diet affect the circulatory system?

12. How can you keep your respiratory system healthy?

Online Health Project

Heart disease is the leading cause of death in America. Find out more about the causes of heart disease. Use the information you find to create a "Healthy Hearts" brochure. You can also create a script for a television or radio that promotes better heart health.

HEALTH LINKS
Go to www.scilinks.org/health.
Enter the code **PMH220** to research **heart disease**.

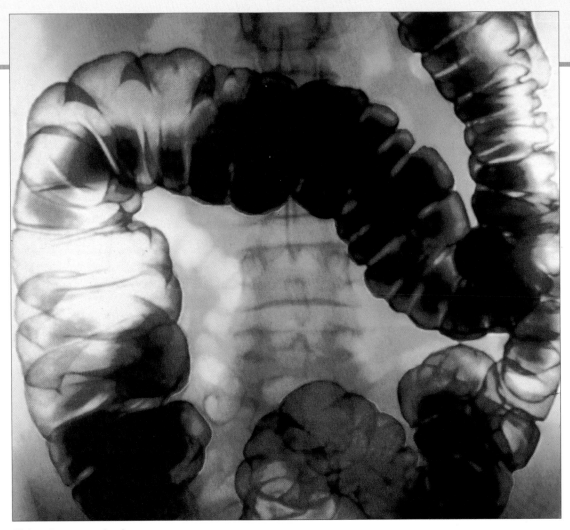

This photograph shows an x-ray of a large intestine. The large intestine is part of your digestive system.

Learning Objectives

- Name the main organs of the digestive and excretory systems and explain what they do.

- Describe some disorders of the digestive and excretory systems.

- List some ways to keep your digestive system and excretory systems healthy.

- **LIFE SKILL**: Explain how you can reduce your risk of food poisoning.

Chapter 4

The Digestive and Excretory Systems

Words to Know

digestion	the process by which the body breaks down food into nutrients that can be absorbed by cells
enzyme	a substance that helps to change chemical reaction rates in the body
saliva	a liquid in the mouth that helps digestion
esophagus	(ih-SAHF-uh-guhs) a tube behind the windpipe that carries food from the mouth to the stomach
pancreas	an organ that produces digestive enzymes and hormones
liver	a large organ that produces bile
bile	a green liquid made by the liver that helps digest fats
gallbladder	an organ that stores bile
nephron	a network of tubes and blood vessels in a kidney that filters waste from blood

Digesting Food

All living things need energy to survive. You get energy from the foods you eat. Your blood carries nutrients from the food you eat all over your body. First, the food must be broken down into simple substances. A whole tuna sandwich cannot move through your blood vessels. The process of breaking down food into nutrients is called **digestion**. This process is carried out by your digestive system.

Talk About It

Why do you think the digestion of food is important to your health?

The Parts of the Digestive System

Talk About It

The appendix is a small finger-shaped structure at the end of the large intestine. An infection of the appendix is called *appendicitis*. Have you or anyone you know had appendicitis? What were the symptoms and treatments?

The digestive system includes many organs. These organs include the mouth, esophagus, stomach, small intestine, and large intestine. They are all connected. In fact, they form one long tube called the digestive tract. The diagram below shows the parts of the digestive system.

In many of these organs, enzymes are involved in digestion. An **enzyme** is a substance that changes the rate of a chemical reaction. When a chemical reaction happens in the digestive system, the process is called *chemical digestion*. Other times, food is broken down into smaller pieces but not changed chemically. This is called *mechanical digestion*. Breaking, crushing, and mashing food are forms of mechanical digestion.

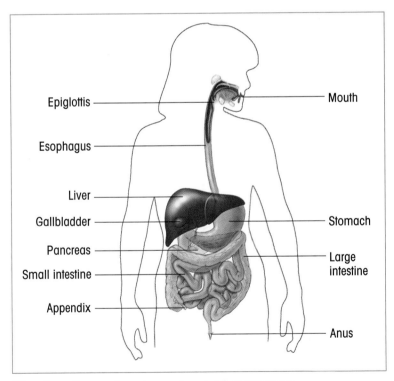

The digestive system is made up of many parts.

The Mouth and the Esophagus

Teeth bite and chew food into smaller pieces. Different teeth are used to do different things. Teeth are covered in a hard material called *enamel*. They also have roots. The *roots* contain nerves and blood vessels. The diagram below shows the four types of teeth and the parts of a tooth.

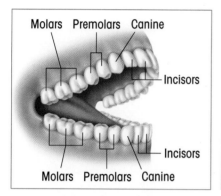

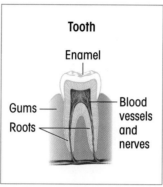

Humans have four types of teeth. A tooth has different parts.

As you chew, a liquid called **saliva** is released. The saliva wets the food and helps you swallow. Saliva also has an enzyme that breaks down starch into sugar. This is a form of chemical digestion.

The **esophagus** connects the mouth to the stomach. Muscles in the esophagus squeeze together, pushing the food down. There is a flap of tissue called the epiglottis between these two openings. The epiglottis closes when you swallow. This keeps food from going down the windpipe.

Health & Safety Tip

Usually, food goes down the correct tube to the stomach. But, food can sometimes go down the windpipe. This causes a person to choke. It is important to chew your food slowly and completely before swallowing.

✓ Check Your Understanding

Write your answers in complete sentences. Use a separate sheet of paper.

1. What are the two types of digestion?

2. What can change the rate of a chemical reaction?

3. What are the four types of teeth?

The Stomach

Food passes from the esophagus into the stomach. It can be stored there for up to six hours. Both mechanical digestion and chemical digestion take place in the stomach. Muscles in the stomach move the food around. Enzymes and acids are released to break down the food chemically. A coating of mucus protects the stomach lining from the acid.

When the food leaves the stomach, it is mostly liquid. The liquid then goes into the small intestine in which most digestion takes place. The small intestine is tightly coiled below the stomach.

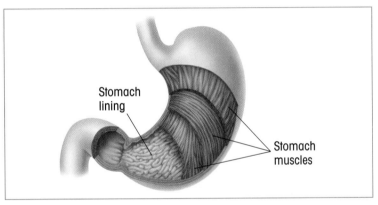

Stomach lining

Stomach muscles

The stomach has muscles that help with digestion.

The Pancreas, Liver, and Gallbladder

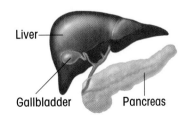

Liver

Gallbladder   Pancreas

The pancreas, liver, and gallbladder take part in chemical digestion.

The pancreas and the liver are two organs that help in the digestion process. The **pancreas** produces enzymes that help digestion. It also produces a substance that neutralizes the acid from the stomach. This neutralizer is important because the small intestine could be damaged by the acid. The **liver** is a large organ that makes a substance called bile. **Bile** is a green liquid that helps digest fat. Bile is stored in the **gallbladder**. The bile is pushed out of the gallbladder into the small intestine.

The Small Intestine and Large Intestine

Next, the liquid food and digestive enzymes move to the small intestine. The lining of the small intestine has many folds and tiny finger-shaped structures called *villi*. They increase the surface area of the small intestine. This means more nutrients can be absorbed. Many tiny blood vessels absorb nutrient molecules into the bloodstream.

You cannot digest all the food you eat. Some leftovers pass into the large intestine or *colon*. The large intestine absorbs water from the leftovers. The water returns to the body. Then, the large intestine forms solid waste. The solid waste is passed out of the body through the anus.

 Check Your Understanding

How do the pancreas, liver, and gallbladder help digestion? Write your answer on a separate sheet of paper.

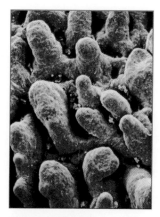

Finger-shaped structures in the small intestine help the body absorb nutrients from food.

Disorders of the Digestive System

An stomachache can be caused by a number of things. You may experience pain in your stomach when you eat too much. Gas in the small intestine can also cause pain. These stomachaches usually go away in an hour or so.

Common Digestive Problems

Sometimes, constipation can cause a stomachache. Constipation occurs when the solid wastes do not move through the large intestine normally. *Laxatives* are medicines used to treat constipation. The best way to get rid of constipation is to eat a lot of fruits, vegetables, and bran. Also, drinking plenty of water and getting some exercise helps. Constipation is usually a sign that you are not eating enough fiber.

Health & Safety Tip

Taking a laxative for a stomachache can be dangerous. If the ache is caused by an infected appendix, a laxative may cause it to burst. A burst appendix can be deadly.

Intestinal flu can cause a stomachache and vomiting. Flu is usually caused by a virus. Most of the time, getting bed rest and drinking fluids are the best ways to treat the flu.

Some forms of bacteria can cause vomiting and diarrhea. If diarrhea or vomiting continues for too long, it can cause a person to become dehydrated. This can be very dangerous. It is important to drink a lot of water if you are sick with vomiting and diarrhea.

More Serious Digestive Disorders

More serious disorders of the digestive system include ulcers, acid reflux, colitis, and even different kinds of cancer. Ulcers are sores in the lining of the stomach or small intestine. Acid reflux occurs when food and acids from the stomach travel back up into the esophagus. This causes a burning sensation sometimes called heartburn. *Colitis* is an inflammation of the colon, or large intestine. Some serious forms of cancer can also occur in the digestive system. These include pancreatic cancer and colon cancer. The best chance for a cure is to find the cancer early. This is done by getting regular checkups and seeing your doctor when you feel sick.

Keeping the Digestive System Healthy

There are a few things you can do to keep your digestive system healthy:

- Try to avoid food poisoning. You can do this by following the steps on page 69.
- Eat foods that are high in fiber. Fiber helps food move more quickly through the intestines.
- Exercise regularly, and get plenty of rest.
- If you have a long-lasting, painful ache, call a doctor. Stay away from antacids and laxatives unless the doctor prescribes them.

✓ **Check Your Understanding**

Write your answers in complete sentences. Use a separate sheet of paper.

1. What are three serious digestive disorders?

2. What causes intestinal flu?

3. CRITICAL THINKING How is fiber helpful to the digestive system?

People in Health

KATIE COURIC—HEALTH ADVOCATE

Katie Couric is the co-host of a popular morning news program. In 1998, her husband, Jay Monahan, died of colon cancer at the age of 42.

Colon cancer was the second deadliest form of cancer in the United States in 2003. It kills over 56,000 people a year. However, this disease is 90 percent curable if it is caught early. Early detection usually involves an examination called a *colonoscopy*. A doctor uses a thin tube with a small camera to look inside the colon. This is not painful, and it takes less than 30 minutes. Many people at risk for colon cancer do not face the problem. They are afraid of getting this exam.

After the death of her husband, Katie Couric decided to do something about colon cancer. She co-founded the National Colorectal Cancer Research Alliance in March 2000. This organization promotes colon cancer research. It also helps to educate people about the disease. In addition, Katie Couric hosted a television special about testing for colon cancer. After the program, there was a 20 percent increase in the number of colonoscopies performed that year.

Katie Couric won an award for her work on colon cancer awareness.

CRITICAL THINKING In what ways is Katie Couric an advocate for better health?

The Excretory System

Wastes are produced as the body carries out life functions. The body needs to get rid of these wastes in order to stay healthy. The excretory system removes wastes from the body. The lungs, kidneys, intestines, and skin are a part of this system.

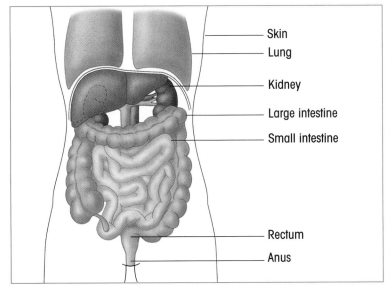

The excretory system removes wastes from the body.

Wastes From Respiration

The lungs get rid of carbon dioxide and water. As you breathe out, you are releasing these waste products into the air.

Solid Waste From Digestion

During digestion, material that cannot be digested and absorbed by the body moves from the small intestine to the large intestine. In the large intestine, water is removed from the material and it becomes solid. The solid waste moves into the rectum. From there, the waste leaves the body through the anus.

Liquid Waste From Digestion

Wastes leave the kidneys as a liquid called *urine*. Urine is made up mostly of water. It also contains salts and other substances. Urine leaves the kidneys through tubes called *ureters*. Then, it is collected in an organ called the *urinary bladder*. Urine leaves the body through another tube called the *urethra*. The drawing below shows these parts of the excretory system.

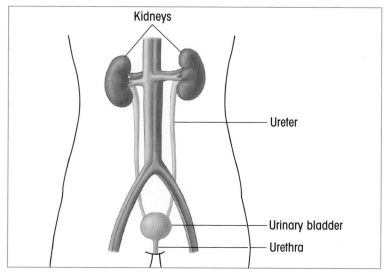

Kidneys

Ureter

Urinary bladder

Urethra

The kidneys, along with the ureter, bladder, and urethra, are parts of the excretory system.

The Kidneys

The kidneys are the organs that remove waste products from the blood. Kidneys are located just above the waistline. These two organs contain a network of small tubes or vessels called **nephrons**. The tubes remove extra water, salt, and other waste materials from the blood.

Each nephron has a cuplike structure at one end. Inside of this structure, blood flows through the many coiled capillaries. Water and other materials are removed from the blood.

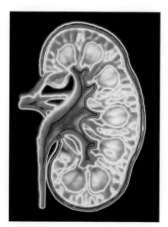

This photograph is a computer image of a kidney.

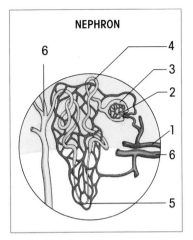

NEPHRON

6
4
3
2
1
6
5

The numbered steps show how this nephron filters wastes and nutrients from blood.

How a Nephron Works

The following steps explain how a nephron filters wastes and nutrients from blood:

STEP 1 Blood enters the nephron in an artery.

STEP 2 The blood passes into coiled capillaries.

STEP 3 Water, salts, and other substances are filtered out. The filtered materials pass into a cuplike structure.

STEP 4 Then, nutrients pass from a tubule back into the blood. Extra water, salts, and waste molecules combine to form urine.

STEP 5 Urine is concentrated in a thin, looped tube. The filtered blood returns to the heart in veins.

STEP 6 The urine then empties into the last part of the nephron called the collecting duct. Urine flows through ureters to the bladder. It is stored there until it is released through the urethra.

Disorders of the Excretory System

Disorders of the excretory system can vary from simple problems to very serious ones.

Cystitis

Normally, your bladder is free of bacteria and germs. However, when bacteria gets into the bladder, you get a bladder infection. This is called *cystitis*. When this happens, there can be a burning sensation every time you urinate. Most of the time, it is necessary to see a doctor when you have cystitis. The doctor may give you antibiotics to clear up the condition. If untreated, bladder infections can spread to your kidneys and cause serious kidney disorders.

Alcohol and the Excretory System

Alcohol causes the kidneys to send more than the usual amount of water out of the body. This water loss leads to *dehydration*, the drying out of your body cells. Too much alcohol can do permanent damage to the kidneys and liver.

Kidney Failure

Kidney failure means that one or both kidneys have shut down. People are able to live normally with only one kidney. But, if both kidneys fail, a person will die if the condition is left untreated.

A *dialysis* machine can be used to treat a person with failed kidneys. The patient's blood is drawn through a tube into a dialysis machine. Waste products and extra fluid in the blood are filtered by the machine. Then, clean blood flows back into the body.

A permanent treatment for kidney failure is a kidney transplant. However, there are never enough "donor" kidneys available.

Talk About It

Do you know anyone who has been on dialysis? How did dialysis treatment help that person? How did it affect his or her daily life?

Keeping the Excretory System Healthy

There are a few ways you can help to keep your excretory system healthy:

- Drink at least six cups of water daily to wash out wastes and bacteria.

- If you are female, wipe yourself from front to back when you go to the bathroom. This will help keep bacteria from getting into your urethra.

- If you think you have a bladder infection, see a doctor.

- To keep your kidneys healthy, stay away from alcohol and caffeine. These drugs can lead to dehydration.

The Skin

The skin covers and protects the body. It also helps rid the body of certain wastes.

The skin contains many sweat glands. Each sweat gland leads to a pore in the skin. The sweat glands release excess water and salts as *perspiration*, or sweat, through the pores.

Perspiration cools down the body. When your body temperature rises, blood circulation increases. Your skin becomes warm. The sweat glands in your skin release sweat. As water in sweat evaporates from your skin, your body cools. Your body temperature lowers.

As this athlete sweats, her body temperature lowers.

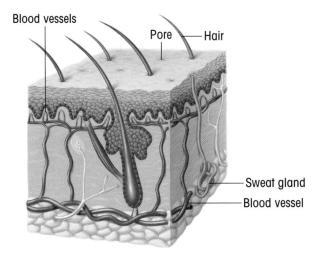

The skin gets rid of wastes through sweat glands.

✓ Check Your Understanding

Write your answers in complete sentences. Use a separate sheet of paper.

1. What organs are part of the excretory system?

2. What are the functions of the kidneys?

3. **CRITICAL THINKING** Which organ in the excretory system reacts to changes in temperature?

LIFE SKILL
Reducing Your Risk of Food Poisoning

You go to a picnic on a hot summer day. After some fun, you eat some chicken and potato salad. A few hours later, your stomach hurts. Intense pain comes in waves, and you feel nauseous. You may also vomit or have diarrhea. Chances are you have food poisoning. Each year, about 76 million people in the United States get sick from spoiled food. Food poisoning comes from eating foods that contain harmful organisms such as bacteria.

You can reduce your risk of food poisoning. Read the tips below. Follow them any time you are preparing or eating food.

Washing your hands before cooking and eating can reduce your risk of food poisoning.

Ways to Avoid Food Poisoning

- Always cook meats thoroughly. If you are not sure what temperature the meat should be when cooked, check the directions on the package.
- Wash fruits and vegetables before eating or cooking them.
- Never eat foods that come from a can that bulges. They may contain harmful bacteria.
- Keep hot foods hot and cold foods cold.
- Always wash your hands, kitchen tools, and counters before, during, and after cooking.

Answer the questions on a separate sheet of paper.
1. What causes food poisoning?
2. What are the symptoms of food poisoning?

Applying the Skill
Create a "Food Safety" poster. Include information about preventing food poisoning. Illustrate your poster with drawings or pictures from magazines.

Summary

Digestion is carried out by many organs, including the mouth, esophagus, stomach, small intestine, and large intestine.

The pancreas, liver, and gallbladder help the digestive process by making enzymes and other digestive fluids.

Disorders of the digestive system include stomachaches, constipation, acid reflux, colitis, and cancer. Good eating habits will help to keep the digestive system healthy.

The excretory system removes wastes from your blood. It does this mainly through the kidneys and skin.

Disorders of the excretory system include bladder infections and kidney failure. Drinking enough water will help keep your excretory system healthy.

Sweat leaves the body through pores of the skin. Sweating releases excess water and salt and cools the body.

bile
digestion
enzyme
esophagus
saliva

Vocabulary Review

Complete each sentence with a term from the list.

1. A tube behind the windpipe that carries food from the mouth to the stomach is called an _____.

2. A substance that helps to change chemical reaction rates in the body is an _____.

3. The process by which the body breaks down food into nutrients that can be absorbed by the cells is called _____.

4. A liquid in the mouth that helps digestion is called _____.

5. A green liquid produced by the liver that helps digest fats is called _____.

Chapter Quiz

**Write your answers on a separate sheet of paper.
Use complete sentences.**

1. What are five organs of the digestive system?

2. What are enzymes?

3. What does saliva do?

4. How do teeth help in digestion?

5. How does food get down the esophagus?

6. What are some ways you can keep your digestive system healthy?

7. What is the job of the excretory system?

8. What is the role of skin in excretion?

9. What is dialysis?

10. What are some disorders of the excretory system?

CRITICAL THINKING

11. Why are villi in the small intestine so important to digestion?

12. Explain how food poisoning can be prevented.

 Online Health Project

In the United States, about 25,000 healthy people die unexpectedly each year. Yet fewer than 20 percent of all Americans choose to become organ donors. Why do you think so few people want to donate their organs? What could be done to change this? Find out more about organ donation. List the steps that are involved in becoming an organ donor. Also, describe how organ donation promotes the health of others. Present your information in a brochure.

HEALTH
LINKS℠
Go to www.scilinks.org/health.
Enter the code **PMH230** to research **organ transplants**.

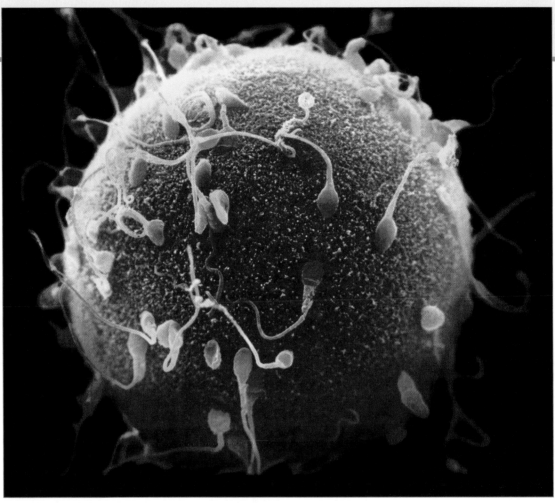

This egg has many sperm trying to fertilize it. Only one will succeed.

Learning Objectives

- Describe the structure and function of the male and female reproductive systems.

- Explain the menstrual cycle.

- Describe how human babies develop inside the mother.

- Describe the stages of human development.

- List and describe some disorders of the male and female reproductive systems.

- **LIFE SKILL**: Explain how you can protect your reproductive health.

Chapter 5 ▶ The Reproductive System

Words to Know

puberty	the time at which a person becomes sexually mature
sperm cell	a male reproductive cell
testes	the male organs that make sperm cells and hormones (singular, testis)
semen	the mixture of fluids in which sperm leaves the body
penis	the male organ that delivers sperm to the female reproductive system
egg cell	a female reproductive cell
ovaries	the female organs that make egg cells and hormones
ovulation	the monthly release of an egg cell from an ovary
Fallopian tube	a tube that leads from an ovary to the uterus
uterus	the female organ in which a fertilized egg develops into a baby
vagina	a canal that leads from a woman's uterus to the outside of her body
menstruation	the monthly shedding of the lining of a woman's uterus
fertilization	the joining of egg and sperm cells
embryo	an organism that is developing from a fertilized egg
placenta	the organ through which nutrients, oxygen, and wastes pass between the mother and the embryo or fetus
fetus	the term used to describe an embryo after 8 weeks of development in the uterus

Human Reproduction

Talk About It

How do you think a new human being is formed?

You have already read about the different organs and systems that make up the human body. In this chapter, you will learn how humans reproduce. You will also learn about how humans develop from one cell into a complete organism.

The Two Sexes

The bodies of men and women work in much the same way. Their hearts pump blood. Their lungs take in air. So, what makes men and women so different?

The main difference between men and women is their reproductive systems. Your reproductive system makes you "male" or "female." Together, the reproductive systems of men and women bring about new life.

While inside the mother's womb, male and female babies look almost the same. At birth, you can tell them apart only by their *genitals*, or external reproductive organs.

Puberty

The human reproductive system is not ready to produce children until puberty. **Puberty** is the time when males and females mature sexually. In males, puberty usually occurs between 13 and 16 years of age. In females, puberty usually occurs between 10 and 14 years of age. However, the age at which puberty begins is different for each individual.

During puberty, visible changes occur in the body. In both men and women, hair starts to grow under the arms and around the genitals. But puberty affects males and females differently. You will learn more about how puberty affects males and females in the next few pages.

The Male Reproductive System

The male reproductive system has two jobs. It must first make sperm cells. **Sperm cells** are male reproductive cells. The male reproductive system also delivers sperm to the female reproductive system.

Sperm cells are made in organs called **testes**. The testes are in a sac of skin. To stay alive, sperm must be kept cooler than other parts of the body. The sac of skin is outside the body so it keeps the sperm cool.

Sperm travel from the testes through thin tubes called *sperm ducts*. These ducts carry the sperm to the urethra. As the sperm move down the sperm ducts and urethra, three glands add fluids to it. The mixture of sperm and fluids is called **semen**. The **penis** delivers the sperm to the female reproductive system.

The reproductive organs in a boy's body begin maturing during puberty. Once puberty begins, the testes make sex hormones. The male sex hormone is *testosterone*. Testosterone causes facial and body hair to grow. It causes boys' bodies to become more muscular. Testosterone also causes boys' voices to become lower.

Remember
Cells are the basic units of structure and function in living things.

The male reproductive system

- Bladder
- Sperm duct
- Urethra
- Penis
- Testes

The Female Reproductive System

Egg cells are the reproductive cells of females. Eggs are made in the **ovaries** of females. The release of an egg is called **ovulation**. First, the egg moves into a tube called a **Fallopian tube**. Tiny hairs push the egg along its path. Then, the egg eventually makes its way to the uterus.

The **uterus** is a muscular, hollow organ. It is where a baby develops if a woman becomes pregnant. The opening to the uterus is the *cervix*. The **vagina** is the canal that leads from a woman's uterus to the outside of her body.

Ovaries also make hormones once puberty begins. The female sex hormones produced in the ovaries are *estrogen* and *progesterone*. Estrogen gives women body hair and broad hips. It also causes their breasts to develop. Progesterone helps the female body prepare the uterus for a baby if she becomes pregnant.

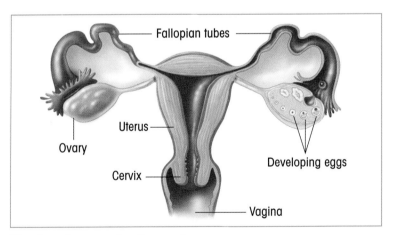

The female reproductive system

 Check Your Understanding

List and describe the parts of the male and female reproductive systems on a separate sheet of paper.

The Menstrual Cycle

In sexually mature women, one or more eggs are released each month. The uterus also changes each month. The walls of the uterus thicken and become swollen with blood. This change prepares the uterus to support a developing baby if the woman becomes pregnant.

Most of the time, the woman does not become pregnant. So, the extra lining of the uterus breaks down. The lining is made of mucus, blood, and dead cells. This material leaves a woman's body through the vagina in a process called **menstruation**.

Menstruation usually lasts from 3 to 6 days. Shortly after menstruation, another egg cell matures in the ovary. Ovulation usually occurs about 14 days after menstruation. Then, in another 14 days, menstruation occurs again. The menstrual cycle repeats monthly.

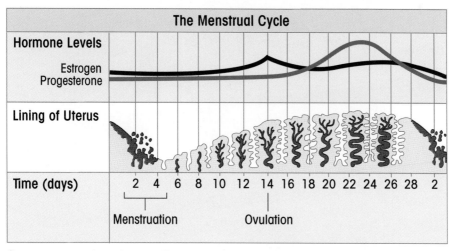

One menstrual cycle usually takes about a month to complete.

Most girls begin menstruating some time between 10 and 14 years of age. Then, some time between the ages of 45 and 55, a woman's menstrual cycle stops happening each month. The permanent end of menstruation is *menopause*.

Pregnancy and Childbirth

When a man and a woman are both sexually mature, they can reproduce, or have a baby. They do so through sexual intercourse. During intercourse, a man inserts his penis into a woman's vagina and releases sperm.

Fertilization

Once released into a female's vagina, sperm cells travel through the cervix and into the uterus. Then, they travel into the Fallopian tubes. **Fertilization**, or the joining of sperm and egg, usually happens in one of the Fallopian tubes. Once the sperm and egg cells have united, a woman becomes pregnant.

After fertilization, the egg begins to divide into many cells. This ball-shaped clump of cells travels to the uterus. The cells attach themselves to the wall of the uterus. The cells are now called an **embryo**.

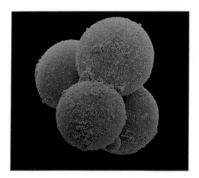 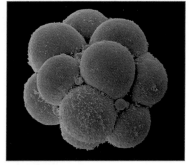

An egg divides into many cells after fertilization.

Remember
Capillaries are tiny blood vessels.

Supporting the Embryo

An organ called the **placenta** forms in the wall of the uterus. Food, oxygen, and wastes pass between the embryo or fetus and the mother through this organ. But, the blood of the mother and of the embryo never mix. The capillaries from both mother and embryo lie very close together in the placenta. A ropelike structure called the *umbilical cord* connects the baby to the placenta.

The Stages of Pregnancy

Pregnancy usually lasts about nine months. Pregnancy is often divided into three periods, or *trimesters*. During each trimester, the embryo will grow in length and in weight. It will also develop organs that it will need to survive outside its mother's uterus. After about eight weeks, the embryo is called a **fetus**. The fetus grows in a pouchlike structure, called an *amniotic sac*.

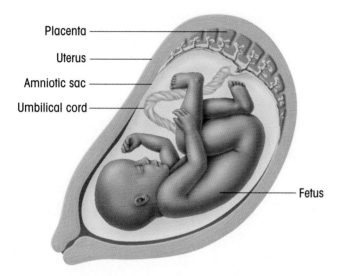

Placenta

Uterus

Amniotic sac

Umbilical cord

Fetus

The fetus grows and develops inside the uterus for about nine months.

Prenatal Care

Prenatal care is the care a baby receives while it is inside its mother. Taking care of her baby before it is born is one of the most important things a mother can do for the baby's health.

Through the placenta, the baby receives food and oxygen from the mother. That means that what goes into the mother's blood goes into the baby's blood. This is why it is important for pregnant women to eat a healthy diet.

Studies have shown that smoking during pregnancy can harm a baby. It can cause babies to be born with low birth weights. These babies have a higher risk of developing illnesses. Drinking alcohol or taking drugs during pregnancy is one of the leading causes of mental retardation in babies. A tiny bit of alcohol is a lot of alcohol to a tiny baby.

Babies born to mothers who drink alcohol while they are pregnant often have a disorder called *fetal alcohol syndrome*. This disorder causes heart defects, brain damage, and physical disabilities. It is best not to drink any alcohol while pregnant. In addition, every pregnant woman should get regular care from a doctor.

Think About It

SHOULD PARENTS HAVE GENETIC TESTING DONE ON THEIR UNBORN CHILDREN?

Improved technology has made it possible to perform medical tests on unborn children. These tests can determine whether the child has an abnormal gene or chromosome. Babies born with these differences may have a birth defect. Cystic fibrosis and hemophilia are birth defects caused by abnormal genes. Down syndrome is a birth defect caused by an abnormal number of chromosomes.

Through genetic testing, parents can learn whether their unborn child has a genetic disease. In some cases, the unborn child can be treated for the disease. Due to the treatment, the child is born healthy. But not all genetic diseases can be treated before birth. Most often, genetic testing simply gives parents information about their unborn child. Then, parents can use this information to make plans for the child's future.

Genetic testing can determine if an unborn baby has traits for certain diseases.

YOU DECIDE Do you think genetic testing on unborn babies should be allowed? Write your answer in complete sentences on a separate sheet of paper. Include reasons for your response.

Childbirth

Shortly before birth, the amniotic sac usually breaks. Strong muscle movements in the mother's uterus, or *contractions*, force the baby out of the mother's body. Contractions are also known as labor pains. The baby leaves the body through the vagina.

After the baby is born, the placenta is also pushed out of the mother's body. Shortly after birth the mother's breasts begin to produce milk. This result is from the effects of hormones from the pregnancy.

The Stages of Human Development

The period before birth is called the prenatal stage. After birth, humans continue to grow and develop. There are five main stages of human development after birth shown in the chart at the right. They are infancy, childhood, adolescence, adulthood, and the later years. The body changes in many ways during each of these stages.

Stages of Human Development	
Stage	Ages
prenatal	before birth
infancy	0–2
childhood	2–12
adolescence	13–18
adulthood	18–55
later years	55+

Disorders of the Reproductive System

One of the most common disorders of the reproductive systems is infertility. *Infertility* is the inability to have children naturally. It may be caused by a physical problem with either the male or the female in a relationship.

Many things can cause a person to be infertile. Men sometimes have a low sperm count. In women, disease or infection can cause scar tissue in the Fallopian tubes. In such cases, the sperm cannot get to the egg. Surgery can sometimes unblock the tubes. But, many women with this problem are infertile. Some women do not produce enough eggs. Finally, some women are born with malformed reproductive organs. This problem, too, can lead to infertility.

 **Health Fact**

Infertility affects about 6 million people in the United States. Some of these couples go through medical fertility treatments in order to have children. Other couples choose to adopt children.

Other problems can occur with the reproductive system. In women, the ovaries can develop masses of tissue called *ovarian cysts*. Ovarian cysts can be harmful to the woman's health. Some women also have irregular menstrual cycles. Irregular menstrual cycles can often be treated with medication.

Cancer can affect the reproductive organs as well. Men can suffer from testicular cancer. Women can have ovarian cancer, breast cancer, and cancer of the uterus. These diseases can sometimes be treated with medication or radiation. But, it is important to detect these cancers early. This can be done through regular doctor's visits and through self-exams.

Sexually transmitted diseases, or STDs, are diseases that are spread through sexual contact. They affect the reproductive organs but can be harmful to other parts of the body as well. Some examples of STDs are AIDS, gonorrhea, and herpes. You will learn more about these diseases in Chapter 7.

Talk About It

Millions of people suffer from STDs. The number of people with STDs is growing every day. What do you think are some ways STDs can be prevented?

✓ Check Your Understanding

Write your answers using complete sentences.

1. On what day of the menstrual cycle does ovulation usually occur?

2. In what organ does a fetus grow and develop?

3. What is it called when a woman is unable to have children naturally?

4. **CRITICAL THINKING** What are some ways a mother can protect the health of her unborn baby?

LIFE SKILL
Protecting Your Reproductive Health

If you are male, you have a one in six chance of getting prostate cancer. Doctors find almost 280,000 new cases of reproductive cancer each year. Reproductive cancers affect men and women. Many cancers can be treated if caught early.

You cannot always avoid all forms of reproductive cancer, but you can take steps to reduce your risk. You can also find cancers early before they become serious. Read the list below on tips for reproductive health.

Common Reproductive Cancers		
Body Part	New Cases (per year)	Deaths (per year)
Breast (in women)	211,300	40,200
Cervix	12,200	4,100
Uterus	40,100	6,800
Ovary	25,400	14,300
Prostate	220,900	28,900
Testes	7,500	400

- **Maintain a healthy lifestyle.** Smoking and being obese are risk factors for many cancers.

- **Make healthy decisions about sex.** The biggest risk factor for cervical cancer is a virus known as HPV. The virus passes between partners during unprotected sex.

- **Learn about screening tests.** Many types of cancer are easier to treat if they are found early. Mammograms are tests that detect breast cancer. The Pap test detects cancer cells in the cervix.

- **Do self-exams.** Boys and men should check their testicles monthly for signs of cancer. Girls and women should do monthly breast exams to check for cancer. Tell your doctor about any lumps.

- **Talk to your doctor.** Discuss your risk factors and ways to reduce them.

Answer the questions on a separate sheet of paper.

1. Which reproductive cancer is most common?

2. Which type is most deadly for women?

Applying the Skill

Choose one reproductive cancer. Create a brochure that answers the following questions: What is it? How common is it? What are the symptoms? How can it be detected in the early stages? How is it treated?

Summary

The testes are male reproductive organs that make sperm cells and hormones.

The ovaries are female reproductive organs that make egg cells and hormones. Egg cells are released during ovulation.

Fertilization usually occurs in the Fallopian tubes. Once the egg and sperm cell join together, a woman is pregnant.

If a woman does not become pregnant, the lining of blood and mucus from the uterus leaves the woman's body during menstruation. Menstruation occurs about once each month.

Pregnancy usually lasts about 9 months. Pregnancy is divided into three periods called trimesters. The health of an unborn baby is affected by actions of the mother.

There are several stages of human development. These include the prenatal stage, infancy, childhood, adolescence, adulthood, and the later years.

Disorders of the reproductive system include infertility, reproductive cancers, and sexually transmitted diseases.

fetus

ovaries

Fallopian tubes

placenta

puberty

sperm

Vocabulary Review

Complete each sentence with a term from the list.

1. Males produce sex cells called _____.

2. The term used to describe offspring after eight weeks of development is _____.

3. Egg cells travel from the ovaries to the uterus by way of the _____.

4. A mother passes food and oxygen to her baby through a tissue called the _____.

5. The female organs that make egg cells and hormones are the _____.

6. The time at which a person becomes sexually mature is called _____.

Chapter Quiz

**Write your answers on a separate sheet of paper.
Use complete sentences.**

1. What is puberty?

2. In which organs are sperm cells made?

3. In which organs are egg cells made?

4. What is the male sex hormone called?

5. What are the two female sex hormones called?

6. Why do blood and mucus leave a woman's body once a month? What is this process called?

7. How does a woman becomes pregnant?

8. How do human babies develop inside the mother?

9. How does a baby receive food and oxygen while inside the mother?

10. What is another term for labor pains?

CRITICAL THINKING

11. In what ways do hormones affect the body during puberty?

12. How can you protect the health of your reproductive system?

Online Health Project

There are many different disorders of the reproductive systems. Use the Internet or other references to find out about one of these disorders. Write a report that describes the causes and treatments for the disorder.

HEALTH LINKS℠

Go to www.scilinks.org/health.
Enter the code **PMH240** to research **reproductive system disorders.**

Unit 1 **Review**

Comprehension Check

On a separate sheet of paper, explain how each of the following prevents disease or keeps your body systems healthy.

1. eating a diet that is low in cholesterol

2. staying away from cigarettes

3. eating a diet high in fiber

4. knowing the signs of reproductive disorders

5. taking the time to exercise regularly

Analyzing Cause and Effect

Write a sentence or two explaining why the following things might happen. The information you learned about in Chapters 1–5 will help you.

6. A person loses muscle control.

7. A person dies of a heart attack.

8. A person has a burning sensation when he urinates.

9. A boy's voice becomes lower.

10. A married couple cannot have a baby.

Writing an Essay

Answer the questions below on a separate sheet of paper. Use complete sentences.

11. How does the nervous system work with your five senses?

12. How are the nervous system and the endocrine system similar? How are they different?

13. How does the respiratory system work with the circulatory system?

14. What is kidney failure and how can it be treated?

15. What changes occur in males and females during puberty?

Managing Your Health

List the ten major body systems. Next to each body system, write two or three things you can do to keep that body system working properly. Present this information in a chart.

Unit 2

Personal Health and Wellness

Before You Read

In this unit, you will learn about your personal health and wellness. You will learn how your body fights disease. You will also learn how you can help prevent disease and injury through your personal actions.

Before you read, ask yourself the following questions:

1. What do I already know about diseases and injuries?

2. What questions do I have about how my body fights disease? How am I protected from injury?

3. How do my actions affect the way my body fights disease, prevents injury, and stays healthy?

Every time you sneeze, you may be releasing disease-causing organisms into the air.

Learning Objectives

- Describe the germ theory of disease.
- Analyze how disease prevention has changed over time.
- Compare how communicable and non-communicable diseases are spread.
- List several forms of pathogens.
- Explain how the body systems fight disease.
- List some ways you can prevent disease.
- **LIFE SKILL:** Access information about flu vaccines.

Chapter **6** ▷ **Fighting Disease**

Words to Know

sterilize	make free of germs, usually by hot water, steam, or chemicals
antibiotic	a medicine that kills harmful bacteria
communicable disease	a disease that is contagious and can be passed from one person to another
non-communicable disease	a disease that is not contagious and cannot be spread by contact among people
epidemic	a widespread occurrence of a certain disease
bacteria	types of one-celled organisms that can cause disease
virus	a microscopic structure that can cause disease
cilia	hairlike structures lining the lungs and airways

The Mystery of Disease

For many centuries, diseases and their causes were mysteries to people. Some people believed that diseases were caused by evil spirits. Others thought they were caused by breathing "bad air."

Today, doctors know that pathogens cause many diseases. For example, we know that germs carried by the fleas on rats caused a great sickness called the bubonic plague. Scientists have learned the causes of many diseases and ways to prevent them. But the causes and cures of other diseases, such as the common cold and cancer, are still not understood.

Write About It

The plague killed millions of people in the 1300s. No one knew what caused it or how to stop it. What do you think it was like to live in those times? How is life better today?

Fighting Disease by Keeping Clean

In hospital rooms of the 1800s, many people died from infections that started in uncovered, or open, wounds. A Scottish doctor named Joseph Lister guessed that germs might be causing these infections. He looked for and found a good "germ killer." It was *carbolic acid*.

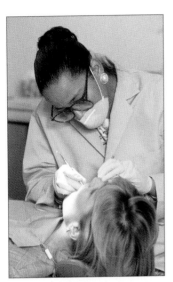

Wearing sterile gloves during medical procedures can help reduce the risk of infection.

Lister began to cover open wounds with bandages soaked in weak carbolic acid. Lister also washed his instruments and his hands with carbolic acid. He even tried spraying his operating room with carbolic acid.

The results were great. The number of infections went down. Many patients stayed alive who would have died before. Lister's carbolic acid was the first *antiseptic*, or germ killer, that worked.

Soon, doctors learned that steam could kill germs, too. Today, medical instruments and bandages are **sterilized** by putting them in hot steam. Doctors wear sterilized gloves, clothes, and masks.

The Germ Theory of Disease

Several years after Lister's discovery of antiseptics, two other scientists found out more about fighting disease. They were Louis Pasteur and Robert Koch. Together these scientists came up with the idea of the *germ theory*. This theory proposed that germs, or microscopic organisms, caused diseases. This theory led to many other advancements in understanding diseases.

Fighting Disease With Medicine

Alexander Fleming was another important scientist. In 1928, he was trying to grow some dangerous bacteria in a small dish. The bacteria were the kind that cause blood poisoning.

One day, Fleming noticed a green mold growing in the dish. It looked something like bread mold. Fleming looked at the mold through a microscope. He saw that a yellow fluid was coming out of the mold. It was killing the bacteria.

Fleming had found a powerful germ killer that could be used as a medicine. He called it *penicillin*. It was the first antibiotic. **Antibiotics** are medicines that are used to kill harmful bacteria. Today, antibiotics save the lives of millions of people suffering from diseases.

✓ **Check Your Understanding**

Write your answers in complete sentences. Use a separate sheet of paper.

1. What is the germ theory?

2. What does it mean to sterilize something?

3. CRITICAL THINKING How was the work of Alexander Fleming important to human health?

People in Health

LOUIS PASTEUR

Louis Pasteur was a French scientist born in 1822. Pasteur helped to develop the germ theory of disease. He also developed a way to control the spread of bacteria in some foods by heating the foods. This method of controlling bacteria is called *pasteurization*.

Pasteur also developed some of the first vaccines. One day, a small boy was bitten by a dog with a disease called rabies. At that time, anyone infected with the disease eventually died. Pasteur had just developed a vaccine for rabies, but it was for animals, not for people. The boy's parents begged Pasteur to try to save their son. He finally agreed to try the vaccine on the boy. Pasteur's vaccine saved the boy's life.

CRITICAL THINKING Explain the importance of Pasteur's work to public health.

Louis Pasteur developed early vaccines.

Understanding Diseases

Understanding the causes of diseases is the first step in preventing them. Diseases can be separated into two groups. These groups are communicable and non-communicable diseases.

Communicable Diseases

A **communicable disease** is one that can be passed from one person to another. Communicable diseases are sometimes called *contagious diseases*. The common cold, influenza or "flu," and strep throat are communicable diseases. AIDS is an example of a more serious communicable disease.

Non-communicable Diseases

A **non-communicable** disease is one that is not contagious. Cancer, high blood pressure, and Lyme disease are non-communicable diseases. These diseases do not spread from one person to another. There are risk factors that can make people more likely to get the disease. These risk factors include smoking, not eating healthy foods, and not wearing insect repellent.

Epidemics

The bubonic plague killed 25 million people in Europe between 1347 and 1400. No one knew what caused the plague or how to stop it.

The plague was called an epidemic because it infected so many people. An **epidemic** is a widespread occurrence of a certain disease. Epidemics still occur today. Many people consider AIDS to be an epidemic. Once started, epidemics are difficult to control.

This photo shows one of the viruses that causes the common cold.

Health Fact

The AIDS epidemic has hit the continent of Africa the hardest. There are more than 25 million people in Africa with HIV, the virus that causes AIDS.

Pathogens

In many cases, diseases start with pathogens in the environment. Many different living and non-living things can be pathogens. Bacteria and viruses are examples of pathogens.

Bacteria are one-celled organisms. Many kinds of bacteria are actually helpful. Bacteria in your large intestine help break down waste matter in your body. But, some bacteria can make you sick. For example, an open cut can become infected by bacteria.

Viruses are microscopic structures that cause disease. Viruses are an unusual kind of pathogen because they are not really alive. However, they can multiply, or reproduce inside living cells. Look at the tables below to find out more about diseases.

Health & Safety Tip

Many pathogens are carried from one place to another by insects or animals. These living things are called *carriers*. Ticks are carriers of Lyme disease. Wearing pants and long sleeves and using insect spray can reduce your risk of getting Lyme disease.

Viral Diseases	
Disease	**Symptoms**
Influenza (flu)	Muscle aches, fever, and chills
Chickenpox	Skin rash in spots and fever
Measles	Pink rash all over the body
Mumps	Swollen glands and fever
Hepatitis	Jaundiced skin and swollen liver

Common Bacterial Diseases		
Disease	**Bacteria**	**Body Part Affected**
Cholera	*Vibrio cholerae*	Small intestine
Lyme disease	*Borrelia burgdorferi*	Skin, joints, heart
Food poisoning	*Salmonella*	Intestine
Strep throat	*Streptococcus pyogenes*	Upper respiratory tract, blood, skin
Tetanus	*Clostridium tetani*	Nerves at synapse

✓ Check Your Understanding

Write your answers in complete sentences.

1. What are two types of disease?

2. What are two kinds of pathogens?

3. CRITICAL THINKING Think about the illnesses you have had this year. What pathogens have you had in your body?

How the Body Fights Disease

Your body has many different ways to fight diseases. It tries to keep the pathogens from getting too far into your body. Different body parts have features that block or kill pathogens. The table below tells about some of your body's defenses.

The Body's Defenses	
Skin	Layers of dead skin cells stop pathogens. Oil in skin helps kill pathogens.
Mouth	Enzymes in the mouth kill pathogens.
Stomach	Acids in the stomach kill pathogens.
Nose	Mucus and hair trap pathogens.

The Respiratory System and Disease

The respiratory system's airways and lungs are lined with cilia. **Cilia** are hairlike structures that are covered with mucus. They move back and forth in a sweeping motion. The mucus on the cilia catches dirt, dust, and germs. The cilia pushes mucus up toward your throat. Then, you swallow the mucus or you spit it out. In this way, the cilia and mucus work together to help keep dirt, dust, and germs out of your lungs. Sneezing is also a way of getting rid of dirt, dust, and pathogens.

The Circulatory System and Disease

White blood cells attack disease pathogens when they enter the body. When your skin is broken, such as when you get a cut, pathogens can enter the body through the cut. White blood cells travel through blood vessels to the area of the cut. Then, they destroy harmful pathogens. Sometimes, a cut will ooze white or yellow liquid. This is called pus. *Pus* is actually a pool of dead white blood cells and other tissues. If pus is present, it means the cut is infected. The diagram on page 95 shows what happens in your body's tissues when you get a cut.

Health & Safety Tip

Always cover your nose and mouth when you sneeze or cough. This will prevent pathogens from being released into the air and spreading to other people.

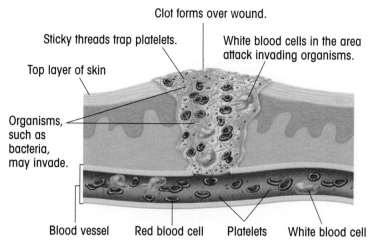

Clot forms over wound.

Sticky threads trap platelets.

White blood cells in the area attack invading organisms.

Top layer of skin

Organisms, such as bacteria, may invade.

Blood vessel   Red blood cell   Platelets   White blood cell

The white blood cells shown here are attacking pathogens that may cause an infection in a cut.

Remember
The circulatory system contains red blood cells and different kinds of white blood cells. White blood cells help the body fight disease. Platelets are pieces of cells that clump together.

The Immune System and Disease

Sometimes pathogens succeed in entering the body. If this happens, the immune system tries to destroy the pathogens. The immune system makes antibodies to fight against them. If the correct antibodies are present when a pathogen enters the body, then the person may not get sick.

Vaccinations are a way to prevent certain diseases. When you were young, you were probably vaccinated against smallpox, measles, and other diseases. When you are vaccinated, your body develops antibodies against the dead or weak germs before the stronger, more serious germs attack.

Scientists have created vaccines for a variety of diseases, including polio, measles, chickenpox, and the flu. There are many more vaccines available today than there were years ago. The development of these new vaccines has meant that many diseases can be prevented. In fact, some diseases no longer exist because of vaccines.

Talk About It

Why do think it is important to keep a record of the vaccinations you have received?

Preventing the Spread of Diseases

Prevent the spread of diseases by following these tips:

- Keep your body clean. Take a bath or shower every day. Also, wash your hands often.
- Keep your house clean. Pathogens grow in uncovered food, dirty toilets, and garbage cans.
- When you are sick, stay home and take care of yourself. You will prevent pathogens from spreading.
- Avoid touching your nose, eyes, and mouth. It will help prevent pathogens from spreading.
- Always cover your nose and mouth when you sneeze or cough. This prevents sending germs into the air.
- Keep your immune system healthy by eating the proper foods, exercising, and getting plenty of sleep.

✓ Check Your Understanding

Write your answers in complete sentences.

1. List three ways to prevent the spread of disease.

2. CRITICAL THINKING Why are vaccinations important to your health?

People in Health

DR. JONAS SALK

During the late 1940s and early 1950s, people lived in fear of a disease called polio. This disease paralyzed, or caused a loss of movement, in some parts of a person's body.

Jonas Salk developed the first vaccine against polio. It was an injection, or a shot. First, he tested the vaccine. In 1955, the vaccine was declared a success. Because of the vaccine, polio was almost completely eliminated from the United States.

CRITICAL THINKING How did Dr. Salk's work change the health of people in the 1950s?

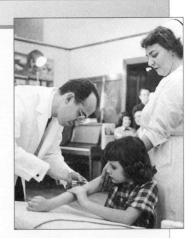

Salk gives a child his vaccine.

LIFE SKILL
Accessing Information About Flu Shots

Have you ever had the flu? Did you have a high fever and feel really tired? You might also have had a really bad headache, a sore throat, a stuffy nose, and achy muscles. Having the flu is not pleasant, but for some people it can be deadly.

Your community may have flu shots available.

A flu vaccine can prevent a lot of the illness and death caused by the flu. Some people, such as babies and older people, are at higher risk of getting very sick. People with certain health problems, such as asthma and HIV, are also at risk. They all should have a flu shot each fall.

Even if you are not in a high-risk group, you still may want to get a flu shot. It can keep you from spreading the virus to high-risk people. Some people should not get the flu shot. People with a severe allergy to hen's eggs should not get it.

Many places offer flu shots. A doctor's office is an obvious choice. Sometimes stores, schools, and health clinics offer flu shots on certain days. Once you decide if you need a flu shot, you can find out where to get one. Here are some sources of information:

FLU SHOT AVAILABILITY

- Your doctor
- The U.S. Centers for Disease Control (http://www.cdc.gov/)
- Your local or state health department

Applying the Skill
Do some research to find out where the flu vaccine is offered in your community. List the places you can go for a flu shot, including where they are located and when they are open.

Answer the questions on a separate sheet of paper.
1. What are some symptoms of the flu?
2. Who may be at high risk?

Summary

Antiseptics, sterilization, and antibiotics are three ways to kill germs.

Diseases can be communicable or non-communicable. A communicable disease is contagious, or may be spread.

Communicable diseases are caused by pathogens. Bacteria and viruses are two kinds of pathogens.

The respiratory system helps the body to fight disease. The nose and cilia help to keep germs out of the body.

The circulatory system has white blood cells that help fight disease by killing germs that cause infection.

The immune system fights diseases by producing antibodies to certain pathogens. These antibodies make a person immune to a disease.

You can prevent disease by keeping your body and home clean and by eating a healthy diet. You can prevent disease from spreading by covering your mouth and nose when you sneeze or cough.

antibiotic

bacteria

communicable disease

non-communicable disease

virus

Vocabulary Review

Complete each sentence with a term from the list.

1. A disease that is not contagious and cannot be spread by contact among people is called a _____.

2. Types of one-celled organisms that can cause disease are _____.

3. A medicine that kills harmful bacteria is called an _____.

4. A disease that is contagious and can be passed from one person to another is called a _____.

5. Microscopic structures that can cause disease are _____.

Chapter Quiz

Answer these questions in complete sentences. Write your answers on a separate sheet of paper.

1. What causes communicable diseases?

2. What does the germ theory say about disease?

3. What is an antibiotic?

4. What is the difference between a communicable and a non-communicable disease?

5. Describe what an epidemic is and give one example.

6. Give one example of a communicable disease and describe how it could spread.

7. What are two kinds of pathogens?

8. List three things you can do daily to help prevent communicable diseases.

CRITICAL THINKING

9. How has the way we prevent diseases changed over the past 200 years?

10. Why is it important to keep your body and home clean?

 Online Health Project

Infectious diseases affect millions of people every day. Some of theses diseases, such as chickenpox and the flu, are fairly common. Other diseases, such as anthrax, SARS, and West Nile virus, are more rare. Research one of these diseases. Write a report that includes the causes, symptoms, treatments, and prevention for the disease. Also, find out what the government is doing to protect people.

HEALTH LINKS.

Go to www.scilinks.org/health. Enter the code **PMH250** to research **infectious disease**.

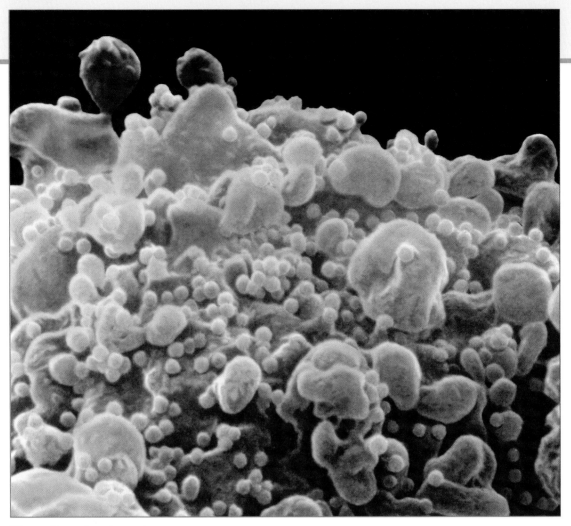

The yellow structures in this photo are the viruses that cause AIDS. They are attached to a white blood cell that is colored pink.

Learning Objectives

- List some examples of the two types of diseases and describe their causes.

- Describe three ways to prevent disease.

- List and describe some inherited diseases.

- **LIFE SKILL:** Analyze how advances in medical technology have influenced human health.

Chapter 7

Diseases and Disorders

Words to Know

AIDS	Acquired Immune Deficiency Syndrome; a disorder that affects the body's ability to fight disease
HIV	Human Immunodeficiency Virus; the virus that causes AIDS
gonorrhea	a sexually transmitted disease that affects the lining of the reproductive organs
chlamydia	(kluh-MIHD-ee-uh) a sexually transmitted disease that affects the vagina in females and the urethra in males
syphilis	a sexually transmitted disease that attacks many parts of the body
genital herpes	a sexually transmitted disease that causes blister-like sores in the genital area
cancer	the uncontrolled abnormal growth of cells
carcinogen	a cancer-causing substance
cardiovascular disease	disorders that affect the heart and blood vessels
diabetes	a disease that affects the way the body changes food into energy
asthma	a type of lung disease that affects breathing
epilepsy	a disease that affects the brain and causes seizures
inherited	passed down from a previous generation in a person's genetic material
sickle cell anemia	an inherited disease that affects the shape and function of red blood cells
hemophilia	an inherited disease that affects the way blood cells clot

Common Communicable Diseases

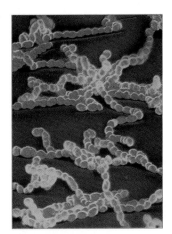

The bacteria shown here cause strep throat.

Health Fact

Some people believe that taking vitamin C pills can cure the common cold. This is not true. There has been no evidence that shows that vitamin C can prevent or cure a cold.

Some examples of common communicable diseases are colds, the flu, and strep throat. Tuberculosis and mononucleosis are also communicable diseases.

Both colds and the flu are caused by viruses. But, they have different symptoms. The symptoms of a cold are runny nose, sore throat, sneezing, and coughing. You can have some of these symptoms with the flu. You may also have headaches, muscle aches, and a fever. There are no medications that can cure the common cold or the flu, but you can take medication to treat the symptoms.

The viruses that cause colds and flu can travel through the air on droplets. The virus is spread when people cough or sneeze without covering their nose and mouth.

Tuberculosis, strep throat, and mononucleosis are examples of other communicable diseases. Each of these diseases is caused by bacteria. *Tuberculosis* is a disease that affects the lungs. *Strep throat* is a more common form of communicable disease. This disease usually causes a sore throat and a fever. *Mononucleosis* is a disease that affects the white blood cells. It is passed by direct contact with an infected person's saliva.

✓ Check Your Understanding

Write your answers in complete sentences. Use a separate sheet of paper.

1. What are three examples of communicable diseases?

2. What causes colds and the flu?

3. CRITICAL THINKING Why are antibiotics not useful in treating colds and the flu?

HIV and AIDS

AIDS stands for Acquired Immune Deficiency Syndrome. **AIDS** is a type of sexually transmitted disease, or STD. A person develops this disease after first getting the **Human Immunodeficiency Virus, or HIV**. This virus infects and destroys cells in the immune system. The immune system protects your body from disease. When the whole immune system fails, the body has trouble fighting off other diseases.

The symptoms of AIDS include weight loss, fevers, rashes, and swollen lymph nodes. Eventually, people with AIDS usually die of diseases that the body can no longer fight, such as pneumonia or cancer. There is no known cure for AIDS.

You can get HIV and AIDS through sexual contact. You can also get the disease by sharing needles to inject drugs into the body. AIDS is also spread from an infected mother to her fetus.

AIDS has become an epidemic. The disease was first recognized by doctors in the early 1980s. By the end of 1991, it was estimated that between 1 and 1.5 million Americans had become infected with HIV. The table below gives information on AIDS cases between 1985 and 2001.

Health Fact

A person can get an HIV blood test. If you are HIV-positive, it means you have the virus. If you are HIV-negative, it means you do not have the virus.

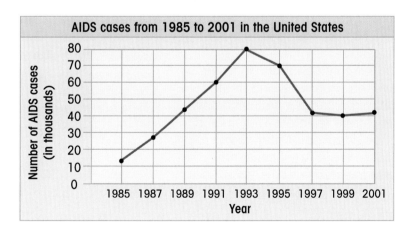

AIDS cases from 1985 to 2001 in the United States

Misconceptions About AIDS

There are a few common misconceptions about AIDS. Some people think that HIV is only passed between same-sex partners. HIV and AIDS can be passed between men and women. HIV is not passed from one person to another through casual contact such as hugging or shaking hands. Also, many people are worried about getting a blood transfusion because they think they may get AIDS. Today all blood used in transfusions is tested for the AIDS virus first. So, getting AIDS through a blood transfusion is very unlikely.

✓ **Check Your Understanding**

Describe the cause and effects of AIDS. Then list the ways AIDS is spread.

Think About It

HOW DO HIV PRIVACY ISSUES AFFECT PUBLIC HEALTH?

Every state has health centers where people can be tested for HIV. Most states have laws requiring these centers to report the names of people who test positive for the virus. The laws also require centers to report the names of people who seek HIV treatment. The purpose of these laws is to keep the public safe by lessening the spread of the HIV virus. However, the laws often stop people with the HIV virus from getting the care they need.

Doctors can give advice about AIDS testing at health centers.

It is estimated that about 280,000 Americans do not know they have the virus because they have not been tested. It is also estimated that only about one in three people who know they have the virus are being treated. They may be afraid of what might happen if others know about the infection. They might be rejected. They might lose their jobs. They might be attacked or hurt. Some people feel that the laws cause more Americans, not fewer Americans, to become infected with the virus.

YOU DECIDE How are state HIV reporting laws helpful to public health? How might they not be helpful?

Other Sexually Transmitted Diseases

Millions of people in the United States suffer from sexually transmitted diseases, or STDs. **Gonorrhea**, **chlamydia**, **syphilis**, and **genital herpes** are four examples of common STDs. The table below gives more information about these diseases.

Sexually Transmitted Diseases	
Disease	**Effects on the Body**
Gonorrhea	burning sensation while urinating; infection in female reproductive system; can cause blindness in babies who get the disease from their mother
Chlamydia	pain and vaginal itching in women; burning sensation in men; can cause men and women to become sterile
Syphilis	open sores on the penis or vagina; can cause blindness, paralysis, and mental illness; can cause death if left untreated
Herpes (two types)	one type causes cold sores on or near your mouth; a second type causes itchy, burning sores on or near the sex organs; can cause fever and more serious illness

How to Protect Yourself From Sexually Transmitted Diseases

STDs are passed through sexual contact. The best way to avoid STDs is not to have sex. Avoiding sexual contact is called *abstinence*. The following is a list of other ways to protect yourself.

- If you are sexually active, make sure that the male partner wears a latex condom. The condom protects both partners from the spread of bacteria and viruses.

- Ask your partner if he or she may be a carrier of an STD. Do not have sex if he or she says yes, or if you have other reasons to believe he or she has an STD.

- Do not put any needles into your skin. Sharing needles can spread the HIV infection and AIDS.

- If you have symptoms of any STD, see a doctor.

Health Fact

Herpes is a very common disease. Seventy-five percent of all people more than five years old have had some form of this virus. Herpes can be treated, but once someone has the herpes virus, he or she has it for life. Herpes symptoms can come and go at any time.

Talk About It

Why is it important for people who test positive for an STD to tell their partner?

Cancer

Cancer is the second leading cause of death in the United States. **Cancer** causes the body's cells to grow abnormally and at an unusually fast pace. Scientists do not know all the causes of cancer. But, studies have shown that contact with substances called carcinogens raises a person's risk of getting cancer. **Carcinogens** are cancer-causing substances. Tobacco, for example, contains carcinogens. Tobacco causes lung, throat, and mouth cancer.

Sometimes, cancer cells gather together into a clump, or *tumor*. These tumors can be either benign or malignant. *Benign tumors* do not spread to other organs. *Malignant tumors* are harmful. They spread to other parts of the body and can damage important organs. Benign tumors are less of a health risk than malignant tumors.

The most common forms of cancer in adults are lung, breast, prostate, and colon cancer. Skin cancer is also common in adults. In children, leukemia is the most common form of cancer. *Leukemia* is a type of cancer that affects the bone marrow and blood.

If you see a red, blotchy mark on your body such as the one shown here, see a doctor.

Some of Cancer's Warning Signs

If cancer is found early, the chances of curing it are quite high. See a doctor if you have any of these signs:

- You have changes in your skin, red patches, or changes in a wart or mole.
- You have any unusual bleeding, especially when you use the bathroom.
- You have a sore or cut that will not heal.
- You have coughing that will not go away.
- You have difficulty with swallowing.
- You have an unusual lump in the breasts or testes. These may be as small as a pea. If you feel anything, see a doctor right away.

Rules for Preventing Cancer

To help prevent cancer, follow these simple rules:

- Stay away from tobacco.
- Eat a diet low in fat and high in fiber.
- Protect your skin from the sun. Wear a hat and use sunscreen.
- If you work around carcinogens, use the proper safety equipment.

✓ **Check Your Understanding**

Write your answers in complete sentences.

1. Name an STD.

2. What is a carcinogen?

3. CRITICAL THINKING How does a person's lifestyle affect his or her risk of getting cancer?

Health and Technology

ADVANCES IN CANCER TREATMENTS

Chemotherapy and radiation are two traditional forms of cancer treatments. *Chemotherapy* is the use of drugs to kill the cancer cells. *Radiation* is the use of radioactive energy to destroy cancer cells. Even though these methods are effective in killing the cancer cells, they may destroy some healthy cells as well. Thus, a person can become weak or very sick.

Many cancer drugs work by stopping body cells from reproducing. In recent years, a new group of drugs has been created to treat cancer. These drugs can actually stop the cancer cells from reproducing, without stopping normal body cells from reproducing. They may prevent the patient from getting very sick after treatment. These new cancer drugs have been created through genetic research.

The black lines on this penny are radiation seeds. They can be implanted into the part of the body that has cancer.

CRITICAL THINKING How have advances in medicine affected the health of people with cancer?

Cardiovascular Disease

Cardiovascular disease is a group of diseases that affect the heart and blood vessels. Many types of cardiovascular disease can be prevented or controlled by a change in lifestyle. For example, a person with high blood pressure can help lower his or her blood pressure by eating healthy foods, exercising, and avoiding salt. A person can reduce his or her risk of having clogged arteries by choosing healthy foods that do not contain too much fat. The table below gives more information about different kinds of cardiovascular disease.

Cardiovascular Diseases		
Disease	**Causes**	**Effect on the Body**
Hypertension (high blood pressure)	obesity, alcohol use, inactivity, smoking, excess salt in diet; age and heredity are also factors	causes heart to work harder; can lead to other cardiovascular diseases
Arteriosclerosis (hardening of arteries)	poor diet that is high in fat, especially cholesterol	buildup of fatty deposits along the inner lining of arteries; arteries become thick and lose elasticity
Atherosclerosis (clogged arteries)	poor diet that is high in fat, especially cholesterol; arteriosclerosis is a contributing factor	buildup of fatty deposits along the inner lining of arteries; arteries become more narrow and eventually completely clogged
Heart Attack	can be caused by clogged or hardened arteries; poor diet, smoking, stress	insufficient oxygen and nutrients get to the heart; heart may beat irregularly or stop beating completely; pain in the center of the chest
Congestive Heart Failure	long period of time with diseased arteries and hypertension; use of illegal drugs	gradual weakening of the heart muscle
Stroke	clogged arteries, hypertension, blood clots	blood supply to the brain is cut off; certain parts of the brain may stop functioning; can affect speech, vision, balance

Diabetes, Asthma, and Epilepsy

Diabetes is the sixth-leading cause of death among Americans. **Diabetes** is a disease that affects the way the body changes food into energy. Diabetes causes the body to not make enough of the hormone insulin. Insulin is needed by the body to change sugar into energy. A low amount of insulin results in high levels of sugar in the blood. This sugar in the blood passes into the urine. From there, it is eventually eliminated, and the body is left without its main source of fuel.

Causes of diabetes include genetic disorders, chemicals, drugs, malnutrition, and obesity. Diabetes is a disease that must be managed throughout a person's life. People who are diabetic usually have to check their blood sugar levels regularly. They also may get injections of hormones.

Asthma is a type of lung disease that affects breathing. It can be life-threatening. Asthma causes breathing problems that often are caused by some trigger in the environment. Large amounts of exercise, colds, allergies, cigarette smoke, and dust or mold in the air can cause an asthma attack to begin. Asthma can be treated by using an asthma inhaler that contains medication. It can also be controlled by avoiding the triggers that lead to an attack.

Epilepsy is a disorder of the nervous system. It is caused by abnormal brain activity. People with epilepsy often have seizures. A *seizure* is when a person loses control of his or her body functions. The person often shakes uncontrollably. Medications can be given that control the seizures. There is no cure for epilepsy.

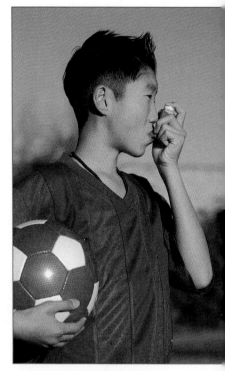

People with asthma may need to take medicine using an inhaler.

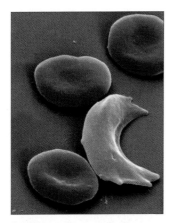

Sickle cell anemia is an inherited disease that causes a person to have some irregular shaped red blood cells.

Inherited Diseases

Some diseases are **inherited**. This means they can be passed from parents to children. **Sickle cell anemia** is an inherited disease. It causes some red blood cells to be misshaped. The disease affects more African American than other groups of people. It can cause weakness and heart problems.

Another inherited disease is hemophilia. **Hemophilia** is a disease that affects the way the blood cells clot. A person with hemophilia may be in danger if he or she gets a cut or a bruise. This person's blood does not clot, and he or she can bleed too much.

There are many other types of inherited diseases. The table below shows some of these.

Inherited Diseases	
Disease	**Effects on the Body**
PKU (phenylketonuria)	body lacks an enzyme needed to break down certain proteins; a buildup of this protein can lead to brain damage and mental disability
Tay-Sachs Disease	body lacks an enzyme that breaks down certain fats; fat can build up in the brain cells and cause death; many people die during childhood
Huntington's Disease	brain cells do not function properly; can lead to loss of muscle control, mental illness, early death
Cystic Fibrosis	causes the respiratory system to produce thick mucus, leading to higher rates of infection by bacteria; can lead to pneumonia and other respiratory illnesses

 Check Your Understanding

Write your answers using complete sentences.

1. What is cardiovascular disease?

2. How can asthma be treated?

3. What causes epilepsy?

4. Name three inherited diseases.

5. CRITICAL THINKING How do your lifestyle choices affect your risk of getting certain diseases?

LIFE SKILL
Analyzing the Influence of Medical Technology

Medical technology has greatly improved our lives. Drugs and devices help us prevent, detect, and fight disease. One example of how technology has improved our health is screening tests. Screening tests are medical tests that help detect problems early. Many diseases are easier to treat if found early.

Heart disease is the leading cause of death in the United States. An *arteriogram* is a type of screening test that shows blocked blood vessels. A doctor and patient can use the information they get from this test to come up with a plan to handle the disease. The knowledge may help a person reduce his or her risk of a heart attack. He or she might become more active, eat better, or take helpful drugs.

A *mammogram* can detect breast cancer. *Prenatal ultrasound* helps monitor the progress of babies in the uterus. Technologies that provide images of brain activity can help people with brain disorders, such as Alzheimer's.

Sometimes the benefit of a new technology is not clear. *Whole body scans* are one example. Some people think these tests will find problems before they have any symptoms. But, so far, there are no proven benefits for healthy people.

Answer the questions on a separate sheet of paper.

1. What is the purpose of screening tests?

2. What type of test can be used to detect breast cancer?

3. **CRITICAL THINKING** Why might getting a screening test for no reason be a bad idea?

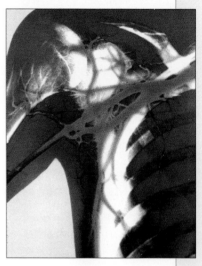

This arteriogram shows some blood vessels of the shoulder and chest.

Applying the Skill
Make a timeline that shows various screening technologies that have been invented over the past 30 years. Describe how each one could be used to protect your health.

Summary

Two forms of disease are communicable and non-communicable. Colds, strep throat, tuberculosis, and AIDS are examples of communicable diseases. Cancer, heart disease, and asthma are examples of non-communicable diseases.

AIDS is a sexually transmitted disease that is fatal. Other sexually transmitted diseases include gonorrhea, syphilis, and herpes. These diseases are spread through sexual contact. The best way to prevent getting an STD is abstinence.

Cancer is the abnormal growth of body cells. Cancers can affect all parts of the body. Some cancers can be prevented by leading a healthy lifestyle.

Cardiovascular disease affects the heart and blood vessels. High blood pressure and heart attacks are examples of cardiovascular disease. Some of these diseases can be prevented by choosing healthy foods that are low in fat.

Diabetes is a disease in which the body does not produce enough of the hormone insulin. Diabetes can often be controlled through diet and medication. Asthma is a disorder of the respiratory system. It makes it harder for a person to breathe. Epilepsy is caused by abnormal brain activity. It can cause a person to have seizures. Some diseases can be inherited.

AIDS

cancer

cardiovascular disease

diabetes

hemophilia

Vocabulary Review

Match the definitions below with the correct term from the list.

1. An uncontrolled abnormal growth of cells is called _____.

2. A disease that affects the way the body changes food into energy is called _____.

3. A disorder that affects the body's ability to fight disease is called _____.

4. A disorder that affects the heart and blood vessels is called _____.

5. An inherited disease that affects the way blood cells clot is called _____.

Chapter Quiz

Answer the questions on a separate sheet of paper. Use complete sentences.

1. What is a communicable disease? Give an example.

2. What is a non-communicable disease? Give an example.

3. What causes colds and the flu?

4. What causes tuberculosis and strep throat?

5. What are the four most common forms of cancer in adults?

6. What are some ways you can reduce your risk of getting cancer?

7. What are three kinds of cardiovascular disease?

8. How can diabetes be treated?

CRITICAL THINKING

9. What are some ways you can avoid getting AIDS?

10. A risk factor is something that makes you more likely to develop a particular disease. What are some of the risk factors for cardiovascular disease?

Online Health Project

Most families today have had one or two members suffer from cancer, cardiovascular disease, or diabetes. Although these diseases are not completely preventable, there are steps you can take to reduce your risk of getting them. Use the Internet to research one of these diseases. Find out what you can do to prevent getting the disease. Make a plan that includes steps to avoid this disease. Include specific behaviors or lifestyle choices in your plan.

HEALTH
LINKS™
Go to www.scilinks.org/health.
Enter the code PMH260 to research **noninfectious disease**.

Wearing a helmet is the best way to prevent a head injury, should you have an accident while skate boarding. There are many other things you can do to prevent injury and stay safe.

Learning Objectives

- Name three ways you can be prepared for an emergency.

- Explain how to prevent injuries.

- Explain how you can stay safe at home, outside, and during activities such as swimming or biking.

- **LIFE SKILL:** Make healthy decisions about safety and injury prevention.

Chapter 8 ▶ Safety and Injury Prevention

Words to Know

fire extinguisher	a metal device filled with water, chemicals, and gases that is used to put out fires
smoke detector	a device that warns people of fire or smoke by letting out a loud noise
hazardous substances	poisonous chemicals that can do harm or kill if swallowed or touched
hazard	any unsafe condition that can cause harm to people
electrical shock	a response to electricity passing through the body
life preserver	a device that helps a person to stay afloat in water
pedestrian	a person who is walking

Preventing Accidents

In the United States, accidents kill almost 100,000 people every year. They cause millions of injuries. You may have times when you think of your environment as unsafe. When you do, it is important to remember that accidents do not "just happen." They are caused.

Most accidents can be prevented. By making certain choices, you can have some control over your environment. You can help to prevent accidents from happening to you and those you love.

Talk About It

What kinds of choices do you think you can make to prevent accidents from happening to you?

Being Prepared

You should always be prepared for an emergency. Being prepared may keep someone from being hurt or from being hurt seriously. Use this checklist as a start:

✓ Keep a list of important telephone numbers.

✓ Make sure you know first aid basics.

✓ Call 9-1-1 if you need to.

✓ Plan an escape route.

Fire

Fires in the home are the third leading cause of accidental death in the United States. Fires can start in ovens. They can also be caused by an old electrical cord or a dried-out Christmas tree. Also, children playing with matches can start fires.

Preventing Fires

Ovens should be fairly clean and free of food and grease. If a fire starts in your frying pan, cover the pan and take it off the burner. Baking soda will help put the fire out. Whatever you do, never put water on a grease fire. Water can cause the fire to spread.

Check your home regularly for other fire hazards. For example, do not plug too many electrical appliances into one socket. If you have a fireplace, make sure you keep the screen closed. Keep matches out of the reach of children.

Emergency Phone Numbers

Mom
696-3498

Aunt Louise
694-1004

Police
876-1294

Fire
876-1296

Poison Control
876-1043

Ambulance
876-9874

Emergency
9-1-1

Make a list like this of your own telephone numbers.

What to Do in Case of Fire

If a fire does start, get everyone out of the house and call the fire department. You should have an escape plan for fire emergencies. If you are trapped in the house, stay close to the floor. Since heat rises, cooler, fresher air is closer to the floor.

If you are in a burning house, put your hand against any door before you open it. If it is hot, look for another way out. If you go into another room, close the door behind you. Open doors allow fire to spread more quickly.

A **fire extinguisher** is a metal device filled with water, chemicals, and gases that is used to put out fire. It is a good idea to keep a fire extinguisher in or near your kitchen. Use it in case of emergencies. If you have a fire in your kitchen, call the fire department quickly.

A **smoke detector** is a device that warns you of fire or smoke by letting out a loud sound. A smoke detector can save your life if a fire does happen. Some smoke detectors also detect *carbon monoxide*, a poisonous gas. You should make sure every floor of your home has at least one smoke detector. Replace the batteries of the smoke detector at least twice a year. It is also a good idea to test the smoke detector periodically to see if it is working properly.

Talk About It

It is a serious crime to pull a fire alarm as a joke. Why do you think it is such a bad thing to do?

Make sure the smoke detectors in your home are working properly.

✓ Check Your Understanding

Write your answers in complete sentences. Use a separate sheet of paper.

1. What should you do if a fire starts in a greasy frying pan?

2. If you are trapped in a burning building, what should you do before opening a door? Why?

3. What is one way you can be prepared for a fire?

Falls

People are injured from falls all the time. Most falls can be prevented. You can do many simple things to prevent falls.

Many people fall down stairs. To prevent falls, make sure the stairs are well lighted. Hold onto the handrails when going up and down the stairs. Do not put anything on the steps. You can trip on even the smallest object placed on a step.

In the kitchen, do not stand on a chair to reach a high shelf. Instead, stand on a sturdy stool or stepladder. Make sure the kitchen floor is dry so you do not slip and fall.

Many people fall in the bathroom. Be sure to put a nonskid mat in the bathtub or shower. It is very easy to slip when your feet are wet. Pick up towels and clothes from the bathroom floor. You can trip over them, too.

People often trip and fall at night when it is dark. Have a light next to your bed. Keep a night light on in hallways and in the bedrooms of small children.

Poisoning

Poisons are things that can hurt you if you swallow them or if they touch your skin or eyes. Poisons can be in foods, plants, medicines, cleaning substances, and chemicals.

Small children are the most likely people to swallow a poison. To prevent this, keep poisons out of the reach of children. Bottles and cans with poisons are often labeled with the words *hazardous substance*. Sometimes the labels tell you what to do in case someone swallows or gets the substance in his or her eyes.

Health & Safety Tip

Never mix household cleaning products. Doing so may create a poisonous gas.

If you or someone you know swallows a poison, call 9-1-1 right away. Many towns and cities have poison control centers that can give you instant advice over the phone. If you can, tell the doctors what was swallowed. Also, note the signs of poisoning. Common signs are coughing, vomiting, and bleeding.

This symbol means the material inside is poisonous.

Electric Shock

We use electricity in our homes every day. We plug in radios, clocks, televisions, kitchen appliances, hair dryers, and lamps. When these objects are not used correctly, they can be a hazard. A **hazard** is any unsafe condition that can cause harm to people.

Electricity comes from an outlet in the wall. A plug is the only thing you should ever put into an outlet. If a small child lives in your home, cover the outlets with special safety plugs.

Sometimes you can get an electrical shock when you plug too many things into the same outlet. An **electrical shock** is a response to electricity passing through the body. Most outlets have room to safely plug in only two things at a time.

Never place an electric appliance in or near water. Even if an appliance is turned off but is still plugged in, electricity is running through it. Never try to open and repair an electric appliance yourself. The damage you may cause can make it unsafe to use.

Here are some other things you can do to prevent an electrical shock:

- Keep all electrical appliances away from water.
- Never touch the metal part of a plug.
- Unplug a lamp before you replace a light bulb.
- Do not use an appliance if the cord is cut.

Health & Safety Tip

Always turn off the electricity at its source when you are working with wiring.

Safety in the Water

To many people, summer is a time for swimming, boating, and water skiing. You can enjoy water activities safely. There are many things you can do to prevent water-related injuries. Here is a list of tips to prevent these injuries:

- Swim only in areas where there is a lifeguard. Most drownings take place where there is no one to help.
- Always swim with a friend. The "buddy system" works. If you are in trouble in the water, your buddy can help you or get help.
- Stay out of deep water if you cannot swim or are a poor swimmer. Do not depend on a rubber float.
- Get out of the water if a storm is coming. If lightning strikes the water, you may be electrocuted.
- Dive in deep water only. Many people have broken their necks from diving into shallow water.
- Throw a rope or floating object to a drowning person. Do not try to save the person yourself unless you have passed a life-saving course.
- Always wear a life preserver while on the water. A **life preserver** is a jacket or other device that will help you to stay afloat in the water.

Wearing a life preserver during boating could help save your life.

✓ Check Your Understanding

Write your answers in complete sentences. Use a separate sheet of paper.

1. What are some ways that you can prevent falls?

2. Where is a good place to store poison?

3. What are some things you can do to prevent an electrical shock?

4. CRITICAL THINKING Why should you only swim in areas where there is a lifeguard?

Safety and Recreation

Many sports and other activities can be dangerous if the proper safety precautions are not taken. Playing sports, skateboarding, using in-line skates, and bicycle riding can all be enjoyed safely if you make smart decisions about safety.

Sports

If you play a sport, you know that injuries cannot always be avoided. But, there are ways to reduce your risk of a sport-related injury. The list of safety tips below will help you avoid such injuries:

- Always wear safety gear, such as pads and helmets.
- Make sure all your equipment works correctly.
- Make sure you know the skills of the sport.
- Follow all of the rules of the sport.
- Do not try to do more than you are capable of doing.

Skateboarding and Skating

Many teenagers enjoy skateboarding and in-line skating. If you are going to take part in these activities, there are a few things you can do to prevent injury. These tips are listed below:

- Wear a helmet, knee pads, elbow pads, and gloves every time you go skateboarding or in-line skating.
- Do not skate in the road where there are cars.
- Do your best to control your speed.
- Go with a friend. In case of an emergency, he or she can get help.

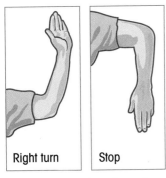

| Right turn | Stop |

Using hand signals makes biking safer for you and those around you.

Bicycle Safety

When you ride a bicycle, remember that you share the road with cars and trucks. These vehicles can cause serious injury or even death if one of them hits you. The most important thing you can do to protect your health while bicycling is to wear a helmet.

Below are some other tips for safe bicycling:

- Do not ride your bicycle on the sidewalk.
- Always ride in the same direction as traffic. Obey all traffic signals signs.
- If you ride with a friend, ride single file.
- Use proper hand signals.
- Make sure your bicycle is in good working condition. Check that there is enough air in the tires.
- Always use a headlight and reflectors at night.

Think About It

SHOULD THERE BE HELMET LAWS FOR BICYCLING?
Today more than 85 million Americans are bicycle riders. Studies show that about 800 bicyclists die in the United States each year. Most of these deaths are caused by brain injuries. Almost 90 percent of these brain injuries would not have happened if the rider had worn a helmet.

17 states now have bicycle helmet laws. They require young riders to wear a helmet when bicycling. Some people feel a law should be passed requiring all riders to wear helmets. Others feel that riders have the right to choose.

YOU DECIDE Should states or towns make it illegal for anyone to ride a bicycle without a helmet? Why or why not?

Many states require you to wear a helmet while bicycling.

Safety on the Road

As a young person, you are going places all the time—work, school, sporting events, and parties. You may be getting there by driving. If you take a driver's education course, you will learn the specific rules for being a safe driver. Here are some general rules for safety on the road:

- Learn defensive driving skills by taking a driver's education course.

- Always follow all traffic laws, especially speed limits.

- Always leave a safe distance between your car and the car in front of you.

- Drive slower in dangerous conditions.

- Wear a seat belt every time you are in a car.

- Make sure every part of your car is working properly.

- Never drive a car after drinking alcohol or taking any drug.

- Never get in a car with a driver who has had alcohol or taken drugs.

Being a safe driver or passenger means wearing a seat belt.

Rules for Safe Walking

╔══════════════════╗

Write About It

Do you think hitchhiking is a safe way to travel? Why or why not? Write your answer on a separate sheet of paper.

Pedestrians, or walkers, are also at risk for injury. Here are some simple rules for safe walking.

- Walk on the sidewalk whenever possible.
- Cross the street only at corners or at pedestrian crossings. Always look before you walk.
- Do not cross if the traffic light has turned red or yellow.
- If you are walking on a road without sidewalks, walk facing the traffic. At night, carry a light.
- Never walk on railroad tracks.

Using Common Sense

Talk About It

Many people, including children, die each year from accidental gunshot wounds. How do you think accidental gunshots can be prevented?

By now, you may be thinking, "This stuff is all common sense. Everyone should know it." Most people do know it, yet they do not take it seriously. The biggest safety rule is this: Think before you act. Whether it is crossing a street or playing a sport, thinking ahead can save your life. Before you go on a camping trip, read up on poisonous plants and animals in that area. If you live in an area where tornadoes or earthquakes occur, learn what to do in these emergencies. Always be careful getting on and off subway cars. Remember that knowledge is power. That means that what you know can save your life.

✓ **Check Your Understanding**

Write your answers on a separate sheet of paper.

1. Name two ways you can prevent injury when playing sports.
2. What is the best way to prevent car accidents?
3. What can you do to be safe while walking?
4. CRITICAL THINKING Why do you think it is important for passengers in the back seat to wear seat belts?

LIFE SKILL
Making Safe Decisions

About 15,000 young people die each year from accidental injury. That equals one death each hour of each day. Many of these injuries can be prevented. Following safety advice can reduce your risk of injury.

You can make decisions to be safe every day. Look at the chart below. It lists a few safety tips. It also shows good reasons for following them.

Safety Tips	
Tips	**Reasons**
Wear a life preserver.	Thousands of people are injured or killed in water-related activities every year.
Wear a seat belt.	Car crashes cause the greatest number of teen injuries. Young people are far less likely to use seat belts than any other age group.
Do not drink and drive.	Alcohol is involved in more than one-third of teen driver deaths.
Wear a helmet when you ride a bicycle, skates, or a skateboard.	Helmets can reduce the risk of a head injury by 85 percent.
Keep any firearms locked and unloaded. Never "play" with a gun.	One in four teen deaths is caused by a firearm.

On a separate sheet of paper, tell what you would do or say in each situation described below. Use complete sentences.

1. You want to go for a bicycle ride. The strap on your helmet is broken.

2. Your friends are going water skiing and they do not want to wear life preservers because they think they look silly.

3. Your friend's father owns a gun. Your friend wants to show it to you.

Applying the Skill

Suppose you are at a party. Your friend has been drinking alcohol. He was supposed to drive you home. What would you do? Suggest different ways to handle the situation. How could you keep yourself safe? How might you convince your friend not to drive the car?

Summary

Accidents injure and kill many people each year. Knowing how accidents are caused is the first step in preventing them.

Some causes of home fires are grease, overcrowded electrical outlets, and fireplaces. Know the escape routes from your home in case of fire. Make sure your home has a fire extinguisher and smoke detectors.

Chemicals and medicines should be kept locked away or out of reach of children. Call a doctor right away in case of poisoning.

Be careful in the water. Swim with a friend. Make sure there is a lifeguard nearby.

Follow safety rules while bicycling, playing sports, and driving in a car.

Use common sense to prevent most accidents from happening.

electrical shock

fire extinguisher

hazard

hazardous substances

pedestrian

smoke detector

Vocabulary Review

Complete each sentence with a term from the list.

1. A device that warns people of fire or smoke by letting out a loud noise is a _____.

2. Poisonous chemicals that can do harm or kill if swallowed or touched are called _____.

3. A person who is walking is called a _____.

4. An unsafe condition that can cause harm to people is called a _____.

5. A metal device filled with water and gases that is used to put out fires is called a _____.

6. A response to electricity passing through the body is called an _____.

Chapter Quiz

Write your answers on a separate sheet of paper. Use complete sentences.

1. Name three ways you can be prepared for an emergency.

2. What are three ways to prevent fires?

3. How can you prevent falls from happening?

4. What are two examples of a hazardous substance?

5. Why should you get out of a pool or lake during an electrical storm?

6. How can you prevent injury while swimming?

7. What is a life preserver?

8. Where is the safest place to cross a street?

CRITICAL THINKING

9. Check your home for safety in the following areas: hazardous substances, electrical danger, fire safety. Describe how your home could be made safer.

10. What can you do to protect the safety of yourself and others while driving?

Online Health Project

Thousands of young people are killed or injured each year in motor vehicle accidents. Many of these deaths and injuries could have been prevented if people took safety precautions more seriously. Find out more about motor vehicle safety. Use the information you find to create a short commercial or skit that demonstrates the ways you can prevent injury and stay safe on the road.

HEALTH LINKS℠

Go to www.scilinks.org/health. Enter the code **PMH270** to research **motor vehicle safety**.

A clean appearance is a sign of good health. Brushing your teeth, washing your face, and wearing clean clothes are some important ways to look good and feel good.

Learning Objectives

- Describe ways to dress appropriately.
- Explain why sleep is important to your health.
- Name three ways to keep your teeth healthy.
- Describe ways to keep your skin, hair, and nails healthy.
- Describe some common problems and treatments for teeth and skin.
- LIFE SKILL: Analyze ways that the media influences people.

Chapter 9 Personal Care and Hygiene

Words to Know

hygiene	taking actions to keep yourself clean and healthy
plaque	a sticky covering on the teeth caused by bacteria
enamel	the outer covering of a tooth
antiperspirant	a product that helps to control perspiration
deodorant	a product that helps control body odors
pimple	a clogged pore that becomes infected
acne	any type of blocked pore
dermatologist	a doctor who specializes in treating skin problems
dandruff	dead skin flaking off the scalp

Personal Care

Taking care of your body is one of the most important things you can do for your health. By taking care of the outside of your body, such as your hair and skin, you are helping to keep the inside of your body healthy. It is important to take care of the inside of your body as well as the outside. By following a few simple steps for personal care and hygiene, you can look good and feel good.

Talk About It

How you feel inside affects how you look on the outside. Why do you think this is true?

Tips for Personal Care

Here are a few tips for taking good care of your body:

- Eat a balanced diet. What you eat makes a big difference in your health and appearance. You will learn more about good nutrition in Chapter 10.

- Exercise regularly. Exercising is good for your physical and emotional health. You will learn more about exercise in Chapter 10.

- Stand and sit straight. Standing and sitting correctly is good *posture*. Slouching makes you look as if you want to hide, as if you do not like yourself.

- Keep your sense organs healthy by protecting your eyes and ears. You read about the ways to protect your sense organs in Chapter 2.

- Dress appropriately.

- Get enough sleep and relaxation.

Dressing Right

What does it mean to dress appropriately? First, it means to wear clothing that is well suited for the weather. Second, dressing appropriately means wearing clothes that are clean. This will keep your skin healthy and reduce odor problems. Third, your clothes should fit well. They should not be too tight or too baggy.

Sleep

Your body cannot function without sleep. Sleep is the way your body recharges itself. Everyone is different, but most people need about 6 to 8 hours of sleep each night. When you are sick, your body needs more sleep to fight the illness.

Wearing the right clothing in winter will protect your body from the cold temperatures and wind.

Many people do not get enough sleep. Studies have shown that half of all adult Americans does not get enough sleep. Lack of sleep can lead to stress, poor work performance, and illness.

How can you tell if you are getting enough sleep? Ask yourself, *Do I wake up feeling good? Do I wake up feeling tired?* If you wake up feeling tired, you are probably not getting enough sleep. Try to go to bed earlier at night.

Besides sleeping at night, it is important to rest and relax from time to time as well. Reading a book, talking with friends, and watching a movie are all healthy forms of relaxation. If you need more rest, taking a short nap may help you feel better.

Health & Safety Tip

If you have trouble falling asleep, try taking a short walk or reading before bedtime. You can also try listening to soft music. If you have trouble sleeping for more than a few nights in a row, talk to your doctor.

Health and Technology

THE SCIENCE OF SLEEP

Snoring and sleep apnea prevent a sound sleep. Snoring is not usually serious. *Sleep apnea* causes a person to stop breathing from several seconds up to two minutes. This can happen as much as ten times in one hour of sleep.

Extra tissue in the upper airways causes snoring and sleep apnea. *Somnoplasty* is a technique used to treat this condition. This technique uses radio waves to shrink the extra tissue. It is performed with a specially designed hand piece that contains a small needle. The needle is used to place the radio waves beneath the extra tissue. In about three to eight weeks, the excess tissue is reduced in size. The decrease in tissue size opens the upper airways. For best results, a patient usually needs more than one treatment.

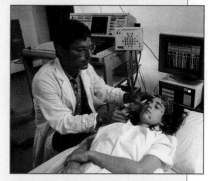

This person is part of a sleep study.

CRITICAL THINKING Why is it important for the upper airways to be open?

Hygiene

Hygiene means taking actions to keep yourself clean and healthy. Brushing your teeth and keeping your body clean are examples of good hygiene.

Taking Care of Your Teeth

One of the most important parts of good hygiene is to have healthy teeth. The mouth is a good place for bacteria to grow. Bacteria live on sugar and bits of food in your mouth. Bacteria can form a sticky covering called **plaque**. If this plaque builds up on your teeth, it can destroy your gum tissue. Left untreated, the plaque will attack the bone that holds your teeth. In the worst cases, plaque can cause your teeth to fall out.

Brushing twice a day and using dental floss once a day will help remove a lot of the plaque. However, it is hard to get all the plaque off by yourself. That is why it is important to see a dentist twice a year. Dentists have special tools for removing plaque.

Teeth have three layers. The first layer is an outer covering called the **enamel**. Cavities are formed when acid made by bacteria eats away at your tooth's enamel. The middle layer is *dentin*. If the acid eats into this layer, you may get a toothache. The next layer is called the pulp. The *pulp* is the core of the tooth. If the cavity reaches the pulp, you may have to have the tooth pulled out. Cavities should be treated before they do lasting damage to your teeth.

Other Common Tooth Problems

If you do not have them already, you will soon be getting wisdom teeth. These teeth come in at the very back of your mouth. Quite often, these teeth will push against other teeth and become infected. In such cases, the wisdom teeth must be pulled.

Health Fact

A survey showed that 4 percent of adults believed that they had gum disease. Actually, 75 percent of adults do.

People sometimes get braces on their teeth during their teenage years. Braces help to adjust teeth so that they fit together properly. Very few people like wearing braces. But, wearing braces for a while can leave a person with a healthy-looking smile. Your dentist can tell you whether these braces are right for you.

Bad breath, or *halitosis*, is caused by bacteria attacking food that is stuck in the teeth. You can prevent bad breath by brushing and flossing regularly. You should also brush your tongue. Rinsing your mouth with an antiseptic may also help prevent bad breath.

Braces help to adjust your teeth into the correct position.

Tips for Healthy Teeth

Follow these simple guidelines to keep your teeth healthy:

- Brush after eating, especially after eating sweets.
- Use dental floss once a day. Flossing gets rid of food between teeth, where many cavities begin.
- Visit a dentist twice a year.
- Remember: All foods can cause plaque!
- Baking soda makes an excellent tooth cleaner and deodorizes the mouth.

Health Fact

If your teeth are discolored or uneven, a dentist maybe able to correct such problems. This work can be very costly.

✓ Check Your Understanding

On a separate sheet of paper, write the words that complete each sentence below.

1. If you do not brush your teeth, _____ will build up on them.

2. Bad breath is caused by _____ in the mouth.

3. Most people need _____ hours of sleep each night.

Taking Care of Your Skin

It is important to wash your face and body with soap and water every day. Washing will remove the dirt and germs that can lead to body odors. Using an antiperspirant or deodorant will also help prevent body odor. **Antiperspirants** reduce the amount of perspiration on your body. **Deodorants** help control body odor.

Caring for your skin also means protecting it from sun damage. You can protect your skin from sun damage by wearing sunscreen. It is a good idea to wear sunscreen with an SPF (sun protection factor) of at least 15. This will help protect your skin from sunburns, wrinkles, and even skin cancer.

Skin Problems

During the teen years, sex hormones bring about a great number of changes in the body. These changes affect the skin. Oil glands in the skin begin to produce more oil. Your face, neck, shoulders, and back may break out with acne. **Acne** is any type of clogged pore. Acne comes in many forms. A whitehead forms when a pore gets clogged with oil, forming a plug. The pore stays white if air cannot reach it. Infection starts when bacteria get under one of these plugs. This is what causes a **pimple**. When air does reach the plug, a chemical change takes place. The plug turns dark. Then, it is called a blackhead.

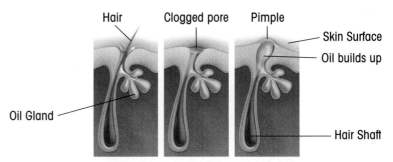

If a pore gets clogged and infected with oil, a pimple will form.

How to Prevent and Treat Acne

Acne is normal for teens. But, it usually gets better as you grow older. Meanwhile, follow these tips:

- Gently wash your face with a mild soap and warm water every day.

- Use over-the-counter medicine for mild acne.

- Keep your fingers away from your face. Squeezing pimples can cause further infection and scars.

- Get plenty of rest and eat fresh, natural foods.

- If you have a bad case of acne, see a dermatologist. A **dermatologist** is a doctor who specializes in skin problems.

Think About It

SHOULD TEENAGERS HAVE COSMETIC SURGERY?
More and more young people today are getting cosmetic surgery. For example, people get the shape of their noses changed. Others get ears that stick out pinned back. Some young women also want to get breast implants.

There are risks that go along with any type of surgery. Cosmetic surgery is no exception. These risks include allergic reactions, infections, or problems with anesthesia. Women who have breast implants may have hardening or lumping of the tissue around the implants, or even bursting of the implant. If the implant bursts, a substance called silicone leaks out. This substance may increase a woman's chance of getting cancer.

Many teenagers are concerned with their appearance.

Most people have some things about their bodies they do not like. Some people think a different nose or fuller lips are worth the money and health risks of surgery. They think getting their bodies "fixed" will improve themselves.

YOU DECIDE Do you think teenagers should be allowed to have cosmetic surgery? Why or why not?

Taking Care of Your Hair and Nails

It is important to keep your hair and nails clean. Oil that is produced by the skin is released onto your hair. Some amount of oil is good for your hair. Too much oil, though, can make your hair feel dirty. Washing your hair with shampoo every day or every other day will keep your hair feeling clean and healthy.

Getting your hair trimmed from time to time will also keep your hair looking healthy. Over time, the ends of your hair break and split. Your hair can look frizzy. Getting haircuts on a regular basis will help prevent this.

Dandruff is a condition that causes white flakes of skin to flake off the scalp. Dandruff is often caused by the extra oil on your scalp. When the oil dries, it causes white flakes. Wash and brush your hair regularly to get rid of dandruff.

Your fingernails and toenails cover and help protect the tips of your fingers and toes. In order to keep your fingernails and toenails from getting too long, you need to trim them. Use a nail cutter or file for trimming. You should also keep your nails and the area underneath your nails clean. You can do this by scrubbing your nails with a nail brush using soap and water. Keeping your nails clean will help prevent the spread of disease.

Health Fact

The body produces ten miles of hair a year!

✓ Check Your Understanding

Write your answers in complete sentences.

1. What are two ways you can keep your skin healthy?

2. What is acne?

3. What causes dandruff?

LIFE SKILL
Analyzing How the Media Influences People

In a magazine, you see an advertisement for a skin cleanser. The model is beautiful. She has perfectly clear skin. If you used the cleanser, would you look like that? The advertiser wants you to think so.

Companies use the media to sell products by selling an image. The term *media* includes newspapers, magazines, radio, television, movies, music, videos, and the Internet. The ads and images in media influence many young people's ideas. Media images affect how people think they should look and act. But, many of the images are not realistic. They suggest that you can achieve this ideal image if you buy certain products.

Unfortunately, most people do not take time to identify why an ad is appealing or if its claims are true. They may simply buy the product because of the image it projects. Do not allow advertisers to manipulate you. Do not let the ads and images in media tell you what you should look like. Try to distinguish facts from fantasy before you buy a product.

Advertisements like this one can change the way you feel about your own appearance.

Look at the advertisement shown above. Then, answer the questions below.

1. What product is the ad trying to sell?

2. What is it about the ad that makes the product appealing?

3. Do you think the product will create the result shown in the ad? Why or why not?

Applying the Skill
What are some ways that you can prevent the media from determining how you should look?

Summary

Eating right, exercising, using good posture, dressing appropriately, and getting enough sleep are all important ways to care for your body.
Bacteria in the mouth cause plaque and cavities. Plaque causes gum and bone disease. Cavities eat into the tooth, causing pain and damage. Brushing and flossing regularly will help keep your teeth healthy.
Acne is caused by clogged, infected pores. Seeing a dermatologist or using over-the-counter medicine is a good way to treat acne.
Washing your hair regularly will keep it looking and feeling healthy. Keeping your nails clean will help you look healthy and prevent the spread of disease.

acne

antiperspirant

dandruff

deodorant

dermatologist

enamel

hygiene

plaque

pimple

Vocabulary Review

Write a term from the list that matches each definition below.

1. Taking actions to keep yourself clean and healthy is called _____.

2. A sticky covering on the teeth caused by bacteria is called _____.

3. A product that helps to control perspiration is called an _____.

4. Any type of clogged pore is called _____.

5. A doctor who specializes in treating skin problems is an _____.

6. Dead skin flaking off from your scalp is called _____.

7. The outer covering of a tooth is called _____.

8. A product that helps control odors is _____.

9. A clogged pore that becomes infected is called a _____.

Chapter Quiz

Answer these questions in complete sentences. Write your answers on a separate sheet of paper.

1. What is one way to dress appropriately?

2. Why is it important to get enough sleep?

3. Why is bacteria likely to grow in the mouth?

4. What will happen if a cavity is left untreated?

5. List three things you can do to keep your teeth healthy.

6. How does a pimple form?

7. Why do many teenagers have acne?

8. Name two ways to treat acne.

9. Name two ways to care for your hair.

10. Why is it important to clean under your fingernails?

CRITICAL THINKING

11. How might your emotional health affect your appearance? Give an example.

12. Predict what might happen if you did not floss your teeth for several weeks.

Online Health Project

Today there are many different treatments and products available for people with acne. Do some research to find out more about these treatments and products. Make a table comparing them. Analyze the usefulness of the treatments or products. Tell whether they have been shown to be effective or if they are just gimmicks.

HEALTH LINKS℠

Go to www.scilinks.org/health. Enter the code **PMH280** to research **acne**.

Eating a variety of healthy foods is a good way to keep your digestive and excretory systems working properly.

Learning Objectives

- Describe the benefits of a balanced diet.
- Identify nutrients in foods.
- Use the Food Guide Pyramid to plan a balanced meal.
- Explain how nutrition and health are related.
- Describe how regular exercise affects your health.
- Define food Calories, and tell how they figure in weight control.
- LIFE SKILL: Design a personal fitness program.

Words to Know

nutrient	a chemical substance the body cannot make but needs to stay alive
cholesterol	a fatlike substance needed by the body in small amounts
Food Guide Pyramid	a diagram that shows the kinds of foods and numbers of servings of each that an average person should eat each day
vitamin	a nutrient found in foods that the body needs and that is made by other organisms
mineral	a nutrient that the body needs and that is found in living and nonliving things
deficiency	a lack of a nutrient
Calorie	a measure of the heat energy found in foods
metabolism	the process by which energy from food is used to carry out life functions

Making Food Choices

Janette and Maria are in the school cafeteria line. Janette puts one piece of chicken, a salad, an apple, and low-fat milk on her plate. Maria picks out a large piece of chocolate cake.

"Hey," says Janette, "no wonder you are gaining weight. Look what you are having for lunch."

"Oh, yeah?" says Maria. "I will have you know that this piece of cake has the same number of calories as your lunch. Also, it is the only thing I am going to eat all day."

Talk About It

Maria is probably right about the number of calories. What do you think about her diet?

You Are What You Eat

"You are what you eat" is an old saying. It does not mean that you will turn into an ice cream cone someday. It means that you will feel and look better if you eat the right kinds of food. Your body, especially when you are growing, needs a *balanced diet*. A balanced diet will help in many ways:

- It will give you energy.
- It will help build new body cells and repair old cells.
- It will help keep your heart and other body organs working well.
- It will help you to think clearly.

Basic Nutrients

Nutrients are the substances your body needs to work well and stay alive. Nutrients are not made by the body. They must be included in your diet. The six kinds of nutrients are water, proteins, fats, carbohydrates, vitamins, and minerals.

Water

Water is an important part of a healthy diet. In fact, about 65 percent of your body weight is water. The cells that make up your body need water to work. Most of the body's chemical processes take place in water. Water helps your body to maintain the proper temperature. It also helps the body get rid of wastes. You should drink six to eight 8-ounce glasses of water each day.

Proteins

No living thing can live without protein. Proteins build muscles, bones, skin, and other solid parts of your body. Proteins can also be used as a source of energy.

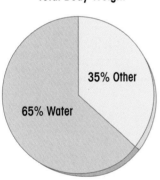

Total Body Weight

35% Other

65% Water

Water makes up a large part of your total body weight.

When you digest the food you eat, the proteins are broken down into smaller parts called *amino acids*. These amino acids help your body build and repair different kinds of cells.

Foods such as milk, eggs, meat, fish, turkey, and chicken have a lot of proteins. Proteins can also be found in soybeans and nuts. Some vegetable protein is in cereal grains and vegetables.

Fats

Your body needs fat to help cushion and protect your organs. Fat also stores energy that you can use in times of need. It also helps to give flavor to the food you eat.

But, too much fat in your diet is unhealthy. Most people already get more than enough fat in their diets. Butter, sour cream, ice cream, bacon, and meat are a few foods that give you animal fat. Margarine, vegetable oils, and nuts contain vegetable oil.

One substance similar to fat is **cholesterol**. Humans need some cholesterol. Cholesterol helps nerve cells to work correctly. It also helps the body to absorb certain vitamins. But, too much cholesterol can lead to clogged blood vessels and heart disease.

Carbohydrates

Carbohydrates, like fats, are "fuel" foods. They give your body at least half of the energy that it needs. There are two forms of carbohydrates—simple and complex.

Simple carbohydrates are sugars. Complex carbohydrates are called *starches*. Your digestive system changes starches and sugars into a sugar called *glucose*. This glucose is used for energy. Simple carbohydrates, or sugars, give your body "quick" energy. They digest more quickly than complex carbohydrates. Complex carbohydrates are found in foods such as rice, bread, crackers, noodles, potatoes, cereals, and vegetables. Complex carbohydrates are better for you than simple carbohydrates.

 **Health Fact**

There are two different kinds of fats—unsaturated and saturated. *Unsaturated fats* include vegetable oils such as canola oil. Butter is an example of a *saturated fat*. Unsaturated fats are healthier for you than saturated fats.

Remember
Proteins and carbohydrates are chemical substances found in foods.

Fiber is a form of carbohydrate that cannot be digested. Although humans cannot digest fiber, it is still an important nutrient. Fiber helps to move food and wastes through the digestive system.

✓ **Check Your Understanding**

Write your answers in complete sentences.

1. What are the six basic nutrients?

2. Why is water an essential nutrient?

3. What are some foods that contain protein?

4. CRITICAL THINKING Suppose you are planning a four-hour hiking trip. What foods or beverages would you take with you that would provide good nutrition during your hike?

Think About It

HOW DO LOW CARBOHYDRATE DIETS AFFECT YOUR HEALTH?

Fad diets are diets that come in and go out of fashion. One popular fad diet is called the low carbohydrate diet. People on this type of diet eat very little or no carbohydrates. Instead, they eat more proteins and fats. A low carbohydrate diet does not allow human body to completely burn the calories found in carbohydrates.

Low carbohydrate diets include meat, dairy, and vegetables.

Research shows that a low carbohydrate diet can bring about weight loss. But this kind of diet is low in calcium and fiber. Also, increased amounts of fat and protein may lead to serious health conditions such as heart disease. Most nutritionists believe that not eating from one food group completely is unhealthy.

YOU DECIDE Do you think a low carbohydrate diet is a good way to lose weight? Why or why not? Write a paragraph explaining your opinion.

Balancing Calories

- Enjoy your food, but eat less.
- Avoid oversized portions.

Foods to Increase

- Make half your plate fruits and vegetables.
- Make at least half your grains whole grains.
- Switch to fat-free or low-fat (1%) milk.

Foods to Reduce

- Compare sodium in foods like soup, bread, and frozen meals and choose the foods with lower numbers.
- Drink water instead of sugary drinks.

Choose**MyPlate**.gov

Fruits

Vegetables

Grains

Protein

Dairy

Vitamins and Minerals

Health & Safety Tip

Do not take vitamins or other supplements without talking to a doctor first. Some vitamins and minerals can be very dangerous if taken in large amounts.

Vitamins are nutrients found in foods that the body needs and that are made by other organisms. Different vitamins serve the body in different ways. All vitamins are important to your health.

Minerals are nutrients that the body needs and that are found in nonliving things. Without minerals, your body could not build healthy red blood cells. The chart below gives you an idea of what some important vitamins and minerals can do for you.

Vitamins and Minerals		
Vitamin/Mineral	**Source**	**Body Function**
A	Leafy green and yellow vegetables, egg yolk, milk, liver, butter, margarine	Good eyesight; healthy skin and hair; growth
B_1	Whole grains, yeast, milk, green vegetables, egg yolk, liver, fish, soybeans, peas	Strong heart, nerves, and muscles; growth; respiration
C	Oranges, grapefruit, lemons, limes, berries, vegetables, tomatoes	Healthy bones; strong blood vessels; helps heal wounds
D	Egg yolk, milk, fresh fish	Strong teeth and bones; growth
Calcium	Milk, vegetables, meats, dried fruits, whole-grain cereals	Healthy bones and teeth; helps blood clotting; prevents muscle spasms
Iron	Liver, meats, eggs, nuts, dried fruits, leafy green vegetables	Formation of red blood cells

 Check Your Understanding

Write your answers in complete sentences.

1. List the six parts of the Food Guide Pyramid.

2. Which vitamin can be found in oranges, grapefruit, and tomatoes?

3. **CRITICAL THINKING** How are vitamins and minerals alike? How are they different?

Reading Nutrition Labels

Most packaged foods made in the United States must include a nutrition label. These labels show the serving size and the number of calories for a serving of that food. They also give other information, such as how much of a certain vitamin the food contains.

Nutrition labels also contain a number called a percent (%) daily value. This information is shown as a percentage of the total daily recommended amount for each nutrient. For example, if a nutrition label read "Vitamin A—10%," that would mean one serving of that food provides 10 percent of the total amount of vitamin A you should be getting each day. Being able to read and understand nutrition labels is an important part of eating healthy. An example of a nutrition label is shown to the right.

Nutrition Facts

Serving Size: 1 cup
Servings Per Container: About 9

Calories 100	
Calories from fat 0	

	% Daily Value
Total fat 0g	0%
Saturated Fat 0g	0%
Cholesterol 0mg	0%
Sodium 10mg	0%
Total Carbohydrate 25g	10%
Dietary Fiber 0g	
Sugars 25g	

How much sugar is found in one serving of this food?

Tips for Good Nutrition

How can you develop good eating habits? Here are a few ideas:

- Drink at least six glasses of water per day.
- Eat a balanced diet.
- Eat snacks only when you are hungry.
- Eat healthy snacks, such as carrots, fruits, and popcorn. Stay away from high-fat snacks, such as potato chips, ice cream, and candy.
- Get into the habit of reading food labels. It will help you to make healthy choices about your nutrition.

✓ **Check Your Understanding**

On a separate sheet of paper, list each of the food groups. Then list two examples from each group.

Write About It

Sometimes people eat when they are nervous or upset. What are some other ways you can deal with your emotions besides eating? Write your answer on a separate sheet of paper.

Nutrition and Health

Nutrition and health are very closely related. What you eat has a direct effect on how you feel and on how your body functions.

People need the right amount of nutrients in order to stay healthy. A **deficiency** is a lack of a nutrient. For example, if a person does not get enough vitamin C, he or she has a vitamin C deficiency. Deficiencies in nutrients can have negative effects on the body. A calcium deficiency can lead to problems with bones and teeth.

Malnutrition occurs when a person does not get enough food overall. This can lead to poor growth or no growth at all. Malnutrition can also cause disease and even death. It is important for your health to eat enough healthy foods every day.

Eating too much of certain foods is also not good for your health. For example, a diet that contains too much fat can lead to serious health problems. These problems include heart disease, high blood pressure, obesity, and diabetes. Eating too much salt can lead to high blood pressure in some people. Eating too much sugar can lead to obesity, which can lead to diabetes. Eating a diet that contains fiber, fruits, and vegetables may lower your risk of certain cancers. These are important things to keep in mind when planning your diet. The table to the left shows how many people are affected by diet-related disorders.

Deaths In Adults	
Causes	**Number**
Heart Disease	700,142
Cancer	553,768
Stroke	163,538
Diabetes	71,372

Many of the leading causes of death may be related to diet.

Fitness

Eating healthy is not the only way to maintain your health. It is also extremely important to stay physically fit through exercise. Physical inactivity is a major risk factor for serious health problems such as heart disease, stroke, and obesity. Many young people do not get as much exercise as they need.

Different Types of Exercise

There are many different forms of exercise. Each type of exercise has different health benefits. Some exercises are designed to increase your flexibility. Other exercises are made to strengthen your bones and muscles. Lastly, there are many forms of exercise that increase your endurance.

Endurance means how long you can continue an exercise. Endurance exercises are especially good for your circulatory and respiratory systems. Playing sports, such as volleyball and soccer, increases your endurance, strength, and flexibility. Many of the activities that you do each day are also good forms of exercise. The exercise pyramid below shows examples of these types of exercises and how often they should be done.

Exercise is an important part of maintaining good health.

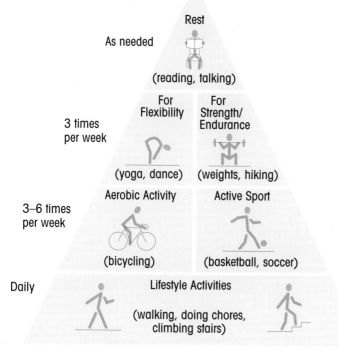

A physical activity pyramid can be used to create a weekly exercise program.

Checking Your Heart Rate

One important skill to use during exercise is to check your heart rate. Your heart rate is the number of times your heart beats per minute. The normal heart rate for a healthy person at rest is about 50 to 80 beats per minute. This rate increases as you exercise. During exercise, your heart rate may increase to more than 150 beats per minute. The calculation below shows you how to determine your heart rate.

Heart rate = number of beats in 10 seconds × 6

A person who is not in good physical condition will have a higher heart rate during exercise than someone who is physically fit. This is because the heart has to work harder to keep up with the physical activity. After exercising regularly for a few weeks, a person's heart rate during exercise will be lower than it was weeks earlier.

Tips for Physical Fitness

Regular exercise will help keep you healthy. Sometimes, people do not know how to find time to exercise. Look at the sample daily planner below to get some ideas.

Sunday	walk to a friend's house
	walk dog
Monday	walk to school
	walk dog
	soccer practice
Tuesday	walk to school
	walk dog
	soccer practice

Including exercise in your daily schedule may be easier than you think.

Fitness and Health

Becoming physically fit through regular exercise is good for your health. Even light exercise two or three times a week has positive effects on your health.

Exercise strengthens your muscles and bones. Having stronger muscles can help lower your risk of being injured during activity. Having stronger bones can help prevent bone diseases such as osteoporosis.

Regular exercise improves the health of your circulatory system, including your heart and blood vessels. Improving the health of your circulatory system reduces the risk of diseases like heart disease and stroke. Exercise has also been shown to lower cholesterol levels in the body. This helps the circulatory system because it lowers the risk of getting clogged blood vessels.

Regular exercise also helps you to maintain a healthy weight. Maintaining a healthy body weight reduces your risk of obesity and diabetes. Finally, exercise can reduce stress and help you live longer.

Write About It

Make a list of ten things you could do to increase your physical fitness.

Talk About It

The number of children and teenagers in the United States who are considered obese has been increasing for the past few decades. What do you think are the causes of this increase?

✔ Check Your Understanding

Write your answers in complete sentences.

1. How can you improve your nutrition?

2. What is a deficiency?

3. List three types of exercise and an example of each.

4. How can you calculate your heart rate?

5. **CRITICAL THINKING** Explain how nutrition and exercise are related to health.

Calories and Your Weight

A food **Calorie** is a measure of heat energy found in food. Different people need different amounts of Calories each day. Your age, body size, and how active you are determine how many Calories you need.

Burning Calories

If you eat more Calories than you burn off, you will gain weight. If you burn off more Calories than you take in, you will lose weight. If the number of Calories you eat equals the amount of energy your body burns, your weight will stay the same. **Metabolism** is the process by which energy from food is used, or "burned," to carry out life functions. Each person burns Calories at a different rate. People with high metabolisms burn Calories quickly. Those with a low metabolism burn Calories slowly. As you grow older, your metabolism will change. Usually, it slows down. This is why older people must be more careful about the foods they eat in order to maintain a healthy weight.

Weight Management

The best way to maintain a healthy weight is to eat the right amounts of a variety of foods and to get regular exercise. The right weight for you depends on your height, your frame, and your body type. If you have concerns about your weight, see a doctor. A doctor can tell you if you really need to lose or gain weight.

✓ Check Your Understanding

Write your answers in complete sentences.

1. What is a Calorie?

2. What is metabolism?

3. CRITICAL THINKING How can you maintain a healthy weight?

Talk About It

Think about your hobbies. Which of these are forms of exercise and which are not? What activities do you think young people commonly engage in instead of exercising? How do these activities affect their health?

LIFE SKILL
Designing a Personal Fitness Program

Are you physically active every day? If you are like many people your age, the answer is probably "No." Studies show that only about one-half of all people ages 12 to 21 are physically active each day.

You can make exercise part of your day. You can set a goal of designing a personal fitness plan. The list below shows the steps to setting a goal.

A. Review the steps below. Write your responses to Steps 1 and 2 on a separate sheet of paper.

STEP 1 Identify the goal you want to reach.

STEP 2 Explain how reaching the goal will affect your health.

STEP 3 Make a plan for reaching your goal. Gather information about different kinds of exercises. Learn about exercises that increase flexibility and those that increase endurance. Discover why all exercise programs begin with warm-up activities and end with cool-down activities.

STEP 4 Evaluate how your plan worked. Are you following your plan? Do you think it is improving your total health? Are there changes that should be made to your plan?

B. Now, use the information in Step 3 to create a personal fitness program. Then, answer the questions in Step 4.

It is important to include exercise as part of your daily routine.

Applying the Skill

After two weeks, you find that your plan is not working. You do not exercise daily. Who could you ask to help you make a new plan?

Summary

A balanced diet can give you energy and a healthy body.

The basic nutrients are water, proteins, fats, carbohydrates, vitamins, and minerals.

The Food Guide Pyramid shows the kinds of foods and numbers of servings of each that an average person should eat each day.

Proper nutrition helps your body to fight off disease.

Exercise can keep you fit and trim for life. The exercise pyramid can help you plan a weekly fitness program.

Foods contain Calories. People burn Calories at different rates. A person can lose or gain weight by controlling Calories. A person's weight depends on a number of factors. If you are unhappy with your weight, see a doctor.

Calorie

cholesterol

Food Guide Pyramid

metabolism

nutrient

vitamin

Vocabulary Review

Use a term from the box to complete each sentence below.

1. A measure of the heat energy found in foods is a _____.

2. The process by which energy from food is used to carry out life functions is called _____.

3. A type of animal fat needed by the body for nerves and digestion is _____.

4. The _____ is a diagram that shows the kinds of foods and numbers of servings of each that an average person should eat each day.

5. An organic substance in food that the body needs to function properly and remain healthy is called a _____.

6. A chemical substance the body cannot make but needs to stay healthy is a _____.

Chapter Quiz

Write your answers on a separate sheet of paper. Use complete sentences.

1. Give three reasons why a person should eat a balanced diet.

2. What are the parts of the Food Guide Pyramid?

3. What are nutrients?

4. If your body needed more protein, what could you eat?

5. How can eating foods with too much fat affect your health?

6. Which vitamins are needed for growth?

7. Give an example of a mineral and what it does.

8. List three forms of exercise, and tell how exercise can affect your health.

CRITICAL THINKING

9. A person eats more calories than he burns off. What will happen to his weight?

10. Why do you think crash diets fail? What is the best way to lose weight? Write at least three sentences on the subject.

Online Health Project

Nutrition supplements are forms of nutrition that are not foods. They often come in the form of pills and shakes. Many more people today are depending on nutrition supplements as a main part of their diet. Find out more about the nutrition supplements that are available today. Analyze the benefits and risks of using such supplements. Summarize your findings in a report.

HEALTH LINKS.

Go to www.scilinks.org/health. Enter the code **PMH290** to research **nutrients**.

Unit 2 **Review**

Comprehension Check
On a separate sheet of paper, write how each of the things below will benefit your health.

1. Wash your hands before you eat.

2. Keep your oven free of food and grease.

3. Swim near a lifeguard.

4. Choose to eat an apple instead of a candy bar.

5. Think before you make a decision.

Analyzing Cause and Effect
Write a sentence or two explaining what might cause the following. Use what you learned in Unit 2 for help.

6. A person who never brushes his teeth has bad breath.

7. A person becomes sick with heart disease.

8. A mother stores cleaning supplies on a high shelf.

9. A whole classroom of students is out sick at the same time with the flu.

10. A boy's face begins to break out in acne when he becomes a teenager.

Writing an Essay
Write an essay to answer each of the following questions.

11. Suppose there is a fire in a building. Why is it important to feel a door in the building before opening it?

12. Pick one infectious and one non-infectious disease to write about. Explain the causes, symptoms, and treatments for each disease.

13. What are two ways you can help stop the spread of disease?

14. Use the Food Guide Pyramid. What is an example of a day's worth of healthy meals?

15. What is an example of a balanced meal, and tell how it will keep you healthy?

Managing Your Health
Write a few paragraphs about what exercise does for your body. Then describe some specific examples of the ways you like to get exercise.

Mental Health

Before You Read

In this unit, you will learn about the brain and mental health. You will learn how your emotional health and stress can affect your overall health. You will also learn about disorders that can affect your mental health.

Before you read, ask yourself the following questions:

1. What do I already know about mental health?

2. What questions do I have about how my emotional health effects my overall health?

3. How should I react to the stresses in my life?

Having a positive outlook on life is a great first step in achieving good mental health. Understanding your personal strengths and weaknesses is another important step.

Learning Objectives

- Name the parts of the brain and describe them.

- Describe how the brain affects behavior.

- Explain how mental and physical health work together.

- Explain how personality is formed.

- **LIFE SKILL:** Describe ways you can access information about mental health resources.

Words to Know

cerebellum	the part of the brain that controls muscle movement
medulla	the part of the brain that controls involuntary actions, such as breathing
cerebrum	the part of the brain that enables us to think, feel emotion, and reason
mental disability	a condition that interferes with a person's ability to think, learn, or speak
mental retardation	a mental disability that limits a person's ability to learn
Alzheimer's disease	an illness that weakens and kills brain cells and destroys the victim's powers of memory and reasoning
emotions	feelings such as fear, anger, sadness, happiness, and love
personality	the combination of many different traits that makes each person unique
temperament	the emotional nature of a person

The Brain

The human brain is not a simple thing to understand. You have already learned how your brain controls your senses, nervous system, and muscles. Scientists and doctors admit there is still much to learn about the brain. What they do know is that the human brain has a lot to do with mental health.

Parts of the Brain

The brain has three main parts: the cerebellum, the brain stem, and the cerebrum. Each part is responsible for different functions.

The **cerebellum** controls the way your muscles work together. Without the cerebellum, you would not be able to coordinate any of your body's movements. You would not even be able to stand up straight.

The brain stem includes the medulla. The **medulla** is the message center of the brain. All messages going to and from the body pass through the medulla. Some of these messages are from your five senses. The medulla also controls your breathing, heartbeat, and digestion. It controls all those bodily functions you do not have to think about.

The **cerebrum** is the largest part of the human brain. The cerebrum enables you to learn, remember, reason, speak, and write. The cerebrum is the "thinking" part of your brain. When people talk about the mind and mental health, they are usually talking about the cerebrum and the activities it controls.

The cerebrum is divided into different sections, or lobes. Each lobe controls different parts of the body and different human behaviors. Look at the drawing below to learn more about the lobes of the brain and the functions that they perform.

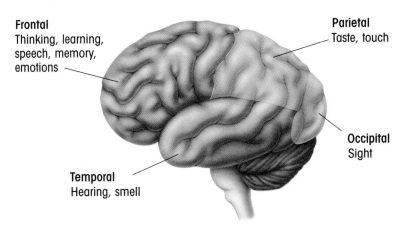

Frontal
Thinking, learning, speech, memory, emotions

Parietal
Taste, touch

Occipital
Sight

Temporal
Hearing, smell

Development of the Brain

The human brain begins developing as early as two weeks into a woman's pregnancy. It begins as a tube made of nerve cells. By about the fifth week of pregnancy, this tube of cells has become bigger, rounder, and more complex. It eventually turns into a small brain.

As the fetus, or unborn baby, develops, the brain also continues to grow and develop. The brain is considered fully formed by about the fourth month of pregnancy. Any harmful substances that the mother takes into her body can damage the developing brain of the fetus.

Brain Damage

Because the brain controls so many functions, damage to the brain can cause serious health problems. Brain damage can impair many human functions, such as motor skills, muscle control, speech, and powers of reasoning, learning, and thinking.

When young people suffer brain damage, it is usually due to accidents. Brain tumors and infectious diseases can also be very damaging or even deadly. Not all tumors are cancerous. Tumors that are cancerous can sometimes be treated with surgery and drugs. *Meningitis* and *chorea* are two infectious diseases that attack the brain.

Remember
A tumor is a mass of cells that grows uncontrollably.

What Are Mental Disabilities?

Mental disabilities are conditions that interfere with a person's ability to think, learn, or speak.

Mentally disabled people can be healthy people. They are not mentally ill. Mental disabilities are caused by birth defects. For example, a person may be born mentally retarded. People with **mental retardation** have a limited ability to learn.

Illnesses or injury can also cause mental disabilities. Strokes or severe head injuries can leave a victim completely paralyzed, unable to speak, think, or learn. Strokes, for example, occur when a blood clot prevents blood from reaching the brain. When that happens, the part of the brain that did not receive enough blood is damaged.

Another illness that affects the brain is **Alzheimer's disease**. This disease strikes older people, causing them to become forgetful. Alzheimer's disease weakens and kills brain cells. Over time, the person loses his or her mental powers of reasoning and becomes confused. There is no cure yet for Alzheimer's.

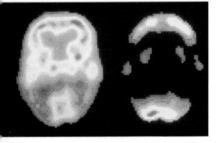

This image shows the brain of a healthy person (left) and the brain of an Alzheimer's patient (right).

✓ **Check Your Understanding**

On a separate sheet of paper, explain how mental health can be affected by changes to the brain.

Health and Technology

BRAIN IMAGING TECHNOLOGY

Brain imaging technology has given researchers a better understanding of how certain parts of the brain function. Two types of technologies used in brain imaging are the Computed Tomograph Scan, or CT Scan, and the Functional Magnetic Resonance Imaging, or fMRI.

CT scans use x-ray beams that pass through the head. The CT scan image shows the structure of the brain. A CT scan does not show how the brain functions.

An fMRI takes a quick series of pictures of the brain. It shows the brain structure. The fMRI also allows doctors to see what the brain is doing when a person is performing certain tasks, such as sleeping, looking at something, or even thinking about something.

Doctors can use imaging technology to study the brain.

CRITICAL THINKING Name at least two things researchers may be able to learn about the brain through brain imaging technology.

What Is Mental Health?

What does it mean to be mentally healthy? Most mentally healthy people like themselves. They like other people most of the time as well.

A mentally healthy person is not happy all the time, of course. But, mentally healthy people do not spend a lot of time being afraid, angry, or jealous. When they do feel these **emotions**, they cope with them in a positive way. People who are mentally healthy have learned ways to control their emotions most of the time.

There are many signs that a person is mentally healthy. A mentally healthy person:

- like himself or herself; has good self-esteem
- knows his or her own strengths and weaknesses
- respects the strengths and weaknesses of other people
- gets along well with others
- has meaningful relationships
- communicates effectively
- makes healthy decisions
- takes responsibility for his or her behavior
- handles stress in healthy ways

Health & Safety Tip

Being a "perfectionist," or expecting too much from yourself, is not good for your mental health. Expecting too little of yourself can also be harmful to your mental health. Try to set realistic goals for yourself.

Mentally healthy people get along with many different people.

Influences on Mental Health

Many factors influence your mental health. These factors include some that you can control and some that you cannot control.

Factors you can control include

- **Your Own Self-image** Your self-image is very important to your mental health. If you have a positive self-image, you will feel happier, and you will get along better with others.

- **Your Physical Health** The physical health of your body affects your mental health in many ways. You will learn more about this relationship later in the chapter.

- **The Relationships in Your Life** By surrounding yourself with positive, supportive people, you can improve your own mental health.

- **The Activities You Take Part In** Certain activities or behaviors are not good for your mental health or your physical health. It is important to understand how the choices you make affect your overall health.

There are many things in life that you cannot control. Sometimes, events happen in your life that you cannot predict. A death in the family and the divorce of parents are two examples. These events affect your mental health. But, you can control how you react to these situations.

Another factor that you cannot control is the way other people feel about you. Not everyone is going to like you all of the time. Many people in the world have different interests and values than you. Some people also have trouble getting along with others. It is better for your own mental health to not let the negative behavior of others make you sad.

Talk About It

How have hurtful comments from others affected your mental health? What are some ways people can replace these negative remarks with positive suggestions?

Maintaining Mental Health

Chapters 12 through 14 will take a closer look at emotional and mental problems and solutions. But for now, here are a few tips for maintaining your mental health:

- **Be positive**. A positive attitude is an important key to good mental health and happiness. When you get up in the morning, tell yourself what a great day you are going to have. If something goes wrong, take a few deep breaths. Think calmly about how to solve the problem, then take action.

- **Change the things you can**. Is there something about your personality or behavior that you would like to change? First, feel good about the things you do like about yourself. Make a list of those things you like about yourself. Then, set a goal for yourself, and take steps to change what you do not like about yourself.

- **Learn to identify and accept the things you cannot change**. Know the difference between something you can control and something you cannot control. Do not waste time worrying about something or arguing over something that is out of your control.

- **Take care of your whole body**. Your physical and mental health affect each other. Practice what you learned in Units 1 and 2 of this textbook to stay healthy.

- **Get help when you need it**. There may be times in your life when you cannot solve your own problems. Perhaps you are sad, lonely, angry, or upset for a long period of time. Learn to recognize these times and go for help. Talk with someone you trust. Find out who the mental health professionals are in your school or community and seek their help.

Health & Safety Tip

Make a list of all the things you like about yourself. Make it a long list! Keep it in your notebook or some other private place. Look at it when you feel badly.

Write About It

Would you have the courage to go to a mental health professional for a problem that was troubling you? Why or why not? Write your answer on a separate sheet of paper.

Body and Mind

The connection between the mind and body is very important. Mental health is very much related to physical health, emotional health, and social health. The health of one affects the health of the other.

But, people can use the connection between body and mind to maintain and improve their health. Some people believe they have helped cure themselves of serious illnesses by "thinking themselves well." They think positive thoughts, and do not let the sickness get them down. These people, and many doctors, believe that a positive attitude can help the body to fight disease.

Think About It

CAN PEOPLE AVOID DEATH?
Robbie's great-grandfather had been sick and dying for months. He held out, though, for his 100th birthday. All his grandchildren and great-grandchildren were there for his party. The next day he died.

Does it seem possible that Robbie's great-grandfather really managed to live through his 100th birthday? A number of researchers have found that people are able to "postpone" death to live through an upcoming holiday or important occasion.

A study was done that examined the death rates of Chinese women over a 25-year period. It concluded that deaths for women over 75 dropped 35 percent the week before the Harvest Moon Festival. Then, the death rate rose by the same amount the week after this important holiday. Why? In Chinese households, the oldest woman leads the activities of the Harvest Moon Festival.

YOU DECIDE Do you think people can use positive thinking to postpone their own deaths? Write your answer on a separate sheet of paper.

Some family members think their older relatives postponed death until after the arrival of a new baby in the family.

Personality

Your personality, like your physical health, has a great deal to do with your mental health. You might say that **personality** is the combination of many different traits that make each person unique. Your personality is a combination of so many different traits—background, interests, values, opinions, goals, abilities, and feelings.

To get a good idea of your whole personality, think about your behavior in different environments, such as school, home, and with friends.

Talk About It

Scientists often find it very useful to study twins who have an identical genetic makeup. Why do you think that is the case?

Personality and Heredity

What makes you different from other people? Your personality is something that is unique to you.

Most scientists agree that heredity plays an important role in shaping your personality. Many of the traits that make up your personality are inherited. They are a mixture of traits from your mother and your father. Your personality may be similar to either parent, but it is still unique to you.

Remember
Inherited traits are those that are passed down to you from your parents.

Some of your traits are inherited from your parents. Your parents inherited some of their traits from their parents.

Your Personality and the Environment

Talk About It

Does your personality change as you move from one environment to another? How?

Some of your personality traits were passed down to you. No one really knows how much of personality is inherited. Environment, attitude, and personal choices also shape your personality. Everyone is born with a certain temperament. Your **temperament** is your emotional nature. A person who is easily excited has an excitable temperament. A person who acts calmly most of the time has a calm temperament.

Just because you are born with a certain temperament does not mean you will always be that way. Your personality and temperament are also shaped by your environment. If a person grows up in a violent household, there is a greater chance that person may become violent later in life. This is sometimes the case with children who are abused by their parents. When abused children grow up, they sometimes abuse their own children. They have learned violent behaviors that became part of their personality.

Write About It

A person's temperament can help or hurt during emergencies. Some people stay calm while others lose control of their emotions. What kind of temperament do you have? Write your answer on a separate sheet of paper.

If behaviors are learned, they can also be "unlearned." With work, new behaviors can take the place of old behaviors. One of the great things about being a human being is that you can learn and change. A shy person can learn to be outgoing. In cases such as child abuse, people can learn to break the cycle of violence.

✓ Check Your Understanding

Write your answers on a separate sheet of paper.

1. What are three ways you can maintain your mental health?

2. How are mental health and physical health related?

3. What is temperament?

4. **CRITICAL THINKING** Explain how a person can take responsibility for his or her personality.

LIFE SKILL
Finding Mental Health Resources

A psychologist will listen to your problems.

There are many places to go for help with mental and emotional health problems. Some problems may require help from a doctor. Other times, talking to people with the same problem can help. The chart below lists some people and places to get help.

Mental Health Resources	
Resource	**Description**
Psychiatrist	Medical doctor who treats mental and emotional illnesses; can prescribe medicines
Psychologist	Has advanced training to diagnose problems and provide therapy
Social worker	Can diagnose problems and provides counseling
Pastoral counselor	Clergy trained to diagnose problems and provide counseling
Mental health clinic	Often provides help at little or no cost
Self-help and support groups	People who help each other get through a problem they have in common

Finding the best source of help may require some research. Often it is a good idea to first talk to your doctor or school counselor. They might be able to suggest the type of person to call.

You can also check your phone book. Often, it has a section that lists mental health resources. Your local mental health clinic or mental health association will also have a list of resources. Finally, an Internet search engine may help you find out about different support groups.

What might you do if you thought a friend needed help from a mental health professional Write your answer on a separate sheet of paper?

Applying the Skill

Use a telephone book or other resource to find out about some mental health resources in your community. Create a poster that lists helpful phone numbers where people can get help.

Summary

The three main parts of the brain are the cerebellum, brain stem, and the cerebrum. The cerebrum is the part of the brain that most people think of as "the mind." The cerebrum enables you to think, feel emotion, speak, write, and remember.

Mental health has to do with behavior, attitude, and emotions. A mentally healthy person likes himself or herself most of the time. A mentally healthy person can cope with emotions and people. Physical health and mental health are closely tied.

A person's personality has a great deal to do with mental health. Personality and behavior can be changed and shaped by the environment and through learning.

You can have a healthy personality by keeping a positive outlook and by learning to like yourself.

In some cases, help from mental health professionals is needed to achieve a "healthy" personality. If you are feeling especially sad, troubled, angry, or upset for a long time, talk to someone about your problems.

Alzheimer's disease

emotions

mental disability

personality

temperament

Vocabulary Review

Match each definition below to a term from the list.

1. a condition that interferes with a person's ability to think, to learn, or to speak _____

2. feelings such as fear, anger, sadness, happiness, and love _____

3. the emotional nature of a person _____

4. an illness that weakens and kills brain cells and destroys the victim's powers of memory and reasoning _____

5. the combination of many different traits that makes everyone unique _____

Chapter Quiz

Answer the questions below on a separate sheet of paper. Use complete sentences.

1. What are the three parts of the brain? What do they do?

2. What is a mental disability? Give two examples.

3. What are two signs that a person is mentally healthy?

4. What factors help shape your personality?

5. Give an example of how mental and physical health work together.

6. A friend of yours gets into fights easily. She says she cannot change. Do you agree, and why?

7. List three factors that influence your mental health.

8. When is it important to get help from a mental health professional?

CRITICAL THINKING

9. Write at least three things that you like about yourself. Is there anything you would like to change about your personality? If so, write a goal and how you will meet it.

10. Suppose someone said, "A mentally healthy person is happy all the time." Would you say this is true or false, and why?

HEALTH LINKS℠

Go to www.scilinks.org/health. Enter the code **PMH300** to research **brain injury**.

Online Health Project

An injury to the brain can have serious effects on a person's mental and physical health. Use the Internet and other resources to find out more about brain injuries. Explain how these injuries affect the body and mind. Describe some possible treatments for brain injuries. Then, list the ways brain injuries can be prevented.

One of the best ways to handle your emotions is to talk about them as soon as possible.

Learning Objectives

- Define emotional health.
- List some common emotions.
- Describe six common types of defense mechanisms.
- Describe how you can handle your emotions in a healthy way.
- Name at least two types of mental health professionals and where they can be found.
- **LIFE SKILL:** Demonstrate positive ways to communicate your emotions.

Words to Know

emotional health	a person's ability to cope with his or her emotions
self-image	how a person feels about himself or herself
defense mechanism	a way of coping with emotions
displaced aggression	taking anger out on a person or thing that did not cause the anger
rationalization	using weak or false reasons to hide the true reasons for bad behavior
compensation	becoming strong in one area to make up for a weakness in another area
daydreaming	a defense mechanism that allows a person to escape through imagination
projection	seeing your emotions in another person
denial	not facing emotions or problems
psychologist	a caregiver trained in giving therapy to people with mild mental disorders
crisis	an extremely difficult situation

When It Gets to Be Too Much

"You are drunk. I do not care if I ever see you again. I never loved you anyway."

Patrick sat in his room and listened to his parents fight. Soon his mother started throwing things. Then, his father angrily left the house. These fights had been going on for years, but they were getting worse.

Patrick could feel his anger increasing. Why did they do this to him? Why were they unable to work things out? Patrick was angry. He decided to call his friend Greg.

"We should play some basketball," said Patrick. "Meet me in the park in half an hour." Greg was there. When he saw Patrick, he rolled up his sleeves.

Two hours on the court and dozens of baskets later, Patrick was worn out. But, he felt better inside. The anger was gone, and he could go home.

What Is Emotional Health?

Talk About It

Some people deal with strong emotions by denying them. Do you think this is healthy?

How you handle anger, fear, and failure determines your emotional health. **Emotional health** is a person's ability to cope with his or her emotions. Usually, emotionally healthy people can cope with emotional stress most of the time.

A person with poor emotional health has trouble handling emotional stress. When the stress is too difficult to cope with, the person may lose control. He or she may try to bury their problems with alcohol, drugs, or negative behavior.

Your Self-image

Emotional health begins with a person's **self-image**. A person with a good self-image can honestly say, "I like myself." Liking yourself means that you are trying to make the most of yourself as a person. It is having self-respect. It means knowing that you are useful as a person, and that others appreciate the things you do.

Liking yourself usually makes others like you. A person with a poor self-image does not like himself or herself. He or she has trouble giving or receiving love.

Understanding Emotions

Emotions are mental and physical changes in your body. It is important to realize that having emotions is not bad. Everyone has them. In fact, sometimes emotions can be helpful to your health. If you feel fear in a dangerous situation, you may react by leaving that situation. This reaction can protect your health.

Other emotions, such as anger, can actually be harmful to your health if you do not deal with them in a positive way. Letting these feelings build up over time can have damaging effects on your physical health. Some studies have shown that people who are angry much of the time may have a higher risk of high blood pressure.

Humans experience a wide variety of emotions—both positive and negative. Teenagers, in particular, can experience more changes in emotions than other people. This is due, in part, to the changes in hormones that a teenager goes through. Below is a list of common human emotions. As you review them, think about the times during your life that you have felt each emotion.

 **Health Fact**

Some fears are reasonable; others are not. Unreasonable fears are called *phobias*. Two common phobias are a fear of heights and a fear of spiders.

Common Emotions		
Emotion	**Explanation**	**Example**
Love	strong affection, concern, and respect	the feeling between a parent and child
Joy	great delight or happiness	the feeling after graduating high school
Empathy	identifying with another person's feeling	feeling sorry for a friend who has been through difficult times
Fear	thought of pain or danger soon to come	feeling scared while walking home alone at night
Worry	anxious or uneasy about something	feeling unsure about how you will do on a test
Sadness	unhappiness or grief	feeling upset after the death of a loved one
Anger	strong displeasure	feeling mad about someone who has bullied you

Defense Mechanisms

How do you cope with stressful emotions? Most people use defense mechanisms. A **defense mechanisms** is a way of thinking or acting that helps relieve your stressful feelings. Defense mechanisms help you keep your self-image and self-respect. But when defense mechanisms are used too often, they can be harmful.

Displaced Aggression

Think about Patrick's behavior. He was angry with his parents. He released his anger on the basketball court. This is called **displaced aggression**. For Patrick, it was a healthy way to deal with his feelings. There was nothing he could do to stop his parents from fighting. So, he turned his energy in another direction.

Displaced aggression is not always the best way to deal with anger. Sometimes it can hurt innocent people or damage property. Suppose Patrick was angry with his girlfriend. It would probably be best for Patrick to let his girlfriend know that he is angry. Then, they could talk about the problem and deal with it. Going to the basketball court might calm Patrick down, but it would not solve the problem. In this sense, displaced aggression would not be a healthy defense mechanism.

Write About It

Talking openly with a good friend or a counselor helps most people to cope with their emotions. List three people that you could talk to openly about your emotions.

Sometimes, sports or other physical activities are good ways to work out your feelings.

Rationalization

A second common defense mechanism is **rationalization**. A person who rationalizes makes up a reason for bad behavior. The person fools himself or herself, or others into believing that he or she really had a good reason for behaving in a foolish or hurtful way. The rationalization removes part of the blame.

Suppose you wake up in the morning, turn over, and then you go back to sleep. When you finally wake up, you are an hour late for school. You tell yourself that it is a good thing you overslept because you do not feel well. You needed the sleep so you would not get sick.

The real reason you did not get out of bed is because you did not want to. When you rationalize, you make yourself feel better for a while. Sometimes, rationalization can keep you from feeling embarrassed in front of others. But, covering up real reasons with false reasons can become a bad habit. It can keep you from dealing with your fears and weaknesses.

Talk About It

Making excuses usually does not fool anybody but you. What do you think are some common things people often make excuses about?

Compensation

A third defense mechanism is compensation. Suppose a girl is not very good at schoolwork. She finds something she is good at—playing the guitar. She works extra hard at her music. This is called compensation. **Compensation** means becoming strong in one area to make up for being weak in another one.

Compensation may be helpful or harmful. A girl who cannot paint well takes up photography. This is helpful. But another girl who does not do well at school tries to compensate by being tough and mean. This is harmful.

In some cases, compensation is not a negative reaction. Focusing on something you are good at can be a healthy behavior.

Daydreaming

Daydreaming is a fourth type of defense mechanism. **Daydreaming** means "escaping your problems by using your imagination." Daydreaming is normal. In most cases, it does not do any harm. Some kinds of "success" daydreaming can even be helpful. People who play sports often imagine themselves winning games or swimming their fastest time. This kind of daydreaming helps them to develop a positive attitude and to perform better.

Sometimes, daydreaming is used to make up for failure. If a person uses daydreaming as an escape, it can become a harmful defense mechanism. The daydreaming robs the person of the time and desire to get something useful done.

Daydreaming is often used as a way to forget about your problems or to put off doing an important task.

Projection

A fifth defense mechanism is called **projection**. When a person projects, he or she imagines that others share his or her same feelings or attitude. For example, an unfriendly person thinks that other people are just as unfriendly. A person who likes to gossip imagines that everyone else also likes to gossip.

Projection, like other defense mechanisms, helps a person maintain self-respect. No one likes to admit to bad qualities. Sometimes projecting bad qualities onto other people makes you feel better about yourself. However, projection can be harmful to others as well as to you. Always thinking of others in a negative way can lead to misunderstandings and sometimes even violence. Besides, boosting your self-respect by thinking less of others does not work over time. In order to have true self-respect, you must be honest with yourself.

Denial

One day, just after gym class, Carol suddenly felt dizzy. Before she knew it, she fell to the floor. The next thing she could remember was her friend Raquel shaking her awake.

"You fainted," Raquel said. "We had better take you to a doctor."

"No," answered Carol. "I am fine. It is nothing."

Carol is using a defense mechanism called denial. She is not accepting, or facing, the facts. **Denial** is not facing emotions or problems. No one likes to be confronted by unhappiness or pain. But, people who carry denial too far can get into serious trouble.

Carol, for example, had not seen a doctor for 3 years. Her fainting spell might have been caused by a number of disorders. It may have been as minor as not having eaten enough that day. It may have been a serious illness that should be treated immediately.

✓ Check Your Understanding

On another sheet of paper, write a definition for each of the following defense mechanisms. Write the definitions in your own words.

1. displaced aggression

2. compensation

3. projection

4. rationalization

5. daydreaming

6. denial

> **H**ealth & **S**afety **T**ip
>
> Try to be as honest with yourself as possible. Denial covers up problems without solving them. It does not make them go away.

Healthy Ways to Handle Your Emotions

Feeling strong emotions does not mean that you are unstable or not emotionally healthy. The way you decide to handle your emotions determines your emotional health.

Defense mechanisms can be healthy if you do not rely on them for too long. Remember, defense mechanisms are a way of avoiding problems. At some time, you will have to face those problems.

There are many positive, constructive ways of dealing with emotions. The list below gives you some tips for dealing with your emotions:

- Examine your feelings. Decide what is really bothering you and focus on that.
- Give yourself time to think about your feelings. Do not react right away.
- Give yourself some space alone.
- Think about whether the problem will seem as bad a few weeks or years from now. This will help you to put your feelings into perspective.
- Write about your feelings in a diary or journal.
- Talk to someone if you need to. Talking about your emotions can help a great deal.
- Realize that your emotions and the things that are causing them may be temporary.
- Remember that everyone goes through emotional problems from time to time.

Where to Get Professional Help

Sometimes talking to a parent, relative, or friend can help you feel better about your emotions. But, if you are not comfortable talking to a person you know, see a good mental health professional.

Here are some places where services are available:

- **Mental Health Clinics** You can find these clinics in your phone book under *mental health,* or under *city* and *county* listings. Very often, you can get help free of charge at these clinics.

- **Church or Medical Doctor** Some church groups offer health and support services. A doctor can refer you to someone who can help with a problem.

- **School** Very often, schools have counselors or psychologists on staff. **Psychologists** are trained professionals who know how to help people with emotional and mental health problems.

Health &
Safety **T**ip

Many communities have hot-line services. Anyone with an emotional problem can call in and get help or advice over the telephone. Some of these services are designed just for people your age.

People in Health

JOE BURNS – GUIDANCE COUNSELOR

Guidance counselors work in schools as advocates for students. They are there to help and support you. Joe Burns has been a guidance counselor for more than 30 years. During a typical day at school, Mr. Burns helps students with many emotional and social problems. Examples of these are family problems or disagreements between friends. He also helps students make positive choices about their education and career planning. Mr. Burns listens to students without judging them. He sometimes refers students to a psychologist or support group for additional help if necessary.

A guidance counselor can help you with problems such as school work or peer pressure.

According to Mr. Burns, some of the most common problems students struggle with are fitting in with their peers, dealing with family issues, and getting good grades. If you are having problems like these or you just need someone to talk to, your school guidance counselor is a good place to start.

CRITICAL THINKING Why do you think it is important for a guidance counselor to refer a student to someone else in certain situations?

When Do You Need Help?

Talk About It

Suppose someone with emotional problems decided to ignore his or her emotions, rather than to seek help. How might this affect his or her health?

Chapter 13 will discuss some more serious emotional disorders. To keep small problems from getting bigger, it is a good idea to talk about your feelings. The following examples are situations for which you should seek professional help:

- When something out of the ordinary happens in your life, it can cause great emotional stress. The death of a loved one, a divorce in the family, or the loss of a friend you really cared about are times of crisis. A **crisis** is a particularly difficult time emotionally. Some signs of crisis are sadness, depression, anger, and withdrawal. You may need to seek the help of a professional during these times.

- If you are using drugs or alcohol, it is a good idea to examine what is going on inside of you. Drugs and alcohol may seem to help you "escape" your problems and emotions temporarily. But, they only add to your problems in the long run. If you are using drugs and alcohol, you may need to discuss your emotions with a professional.

- If your emotions prevent you from eating, sleeping, or carrying out other daily activities, you need to seek help. If you feel you just cannot get out of bed because of your emotions, you need to seek help.

✓ Check Your Understanding

Write your answers in complete sentences.

1. What are three ways to handle your emotions?

2. List two types of professionals who can help you with your emotions.

3. What are two examples of a crisis?

4. CRITICAL THINKING Why do you think it is important for someone who is using drugs and alcohol to seek professional help?

LIFE SKILL
Communicating Emotions in Healthy Ways

A good friend has a bad habit. You always listen to her, but it feels like she rarely listens to you. It seems like her ideas and problems are more important than yours. What should you do?

One choice is to let her interrupt you. But, you might feel more hurt. When you do speak up, you may be angry.

The best approach is to communicate your emotions in a positive way. Using "I messages" can help you communicate effectively. With "I messages," you explain how you feel and why.

Parts of an "I Message."

"I feel..." (describe how you feel, such as angry, hurt, or worried)

"when you..." (explain what is bothering you)

"because..." (tell why this bothers you)

By following a few tips, you can communicate your emotions in a positive way.

- Communicate with confidence. Use "I messages." Tell the person exactly what behaviors you do and do not like.

- Avoid attacking messages. A verbal attack can lead to hard feelings.

- Be a good listener.

- Be aware of angry feelings. Take time to calm down before telling how you feel.

You studied hard for a test. You realize a friend copied your answers. How would you feel? Write a dialogue between you and your friend.

Applying the Skill

Write an "I message" to each person below.
1. Your younger brother lost your newest CD.
2. A friend puts down your basketball team because it has lost a few games.
3. Your mother walks into your room without knocking.

Summary

Emotional health is measured by how well a person deals with feelings or emotions. An emotionally healthy person can face stressful situations and find positive ways to cope with them.

Emotional health begins with a good self-image. Put simply, people with good self-images like both their bodies and personalities.

People often use defense mechanisms to cope with their emotions. Defense mechanisms can be helpful or harmful, depending on how they are used.

Healthy ways to handle your emotions include examining your feelings, writing them down, talking with someone, and getting professional help when you need it.

Crises are particularly difficult times emotionally. It is important to recognize the signs of crisis in yourself and in your loved ones.

crisis

defense mechanisms

emotional health

psychologist

rationalization

self-image

Vocabulary Review

Complete each sentence below with the correct term from the list.

1. The ways of coping with emotions are called _____.

2. Using weak or false reasons to hide the true reason for bad behavior is called _____.

3. A person's ability to cope with his or her emotions is _____.

4. How a person feels about himself or herself is called _____.

5. A caregiver trained in giving therapy to people with mild mental disorders is a _____.

6. An extremely difficult situation is called a _____.

Chapter Quiz

Write your answers on a separate sheet of paper. Use complete sentences.

1. What does it mean to have good emotional health?

2. Describe a person who has a good self-image.

3. How does the use of defense mechanisms help to keep you emotionally healthy?

4. How can defense mechanisms be harmful?

5. Give an example of how daydreaming can be healthy.

6. A friend drinks alcohol every day at lunchtime. He says he can quit any time he wants to. What defense mechanism is he using? Is it helpful or harmful?

7. Name three places or ways you can find mental health professionals.

8. Give an example of a crisis situation.

CRITICAL THINKING

9. Think back to a time when you felt emotional stress. What steps could you have taken to make yourself feel better?

10. How could you promote good emotional health in others? List several ways.

Online Health Project

Most of us get angry from time to time. This is a normal emotion. However, some people react in very negative ways to their anger. They may become hostile or even abusive when they are angry. How can anger be managed before it gets to this point? Research some anger management techniques. Think about which techniques seem most useful for you and your personality. Then, summarize the technique in a poster or brochure.

HEALTH LINKS.

Go to www.scilinks.org/health. Enter the code PMH310 to research **anger management**.

It is good to talk your problems over with someone when you feel upset.

Learning Objectives

- Name the kinds of mental disorders.
- Describe some forms of neuroses and psychoses.
- List the signs of depression.
- Describe warning signs of suicide.
- Compare eating disorders.
- LIFE SKILL: Demonstrate how to raise awareness about preventing teen suicide.

Words to Know

mental disorder	a condition that disturbs a person's emotions, thinking, and behavior
neurosis	a mild type of mental disorder that may be treated with therapy
anxiety	a deep fear or worry that something bad is going to happen
obsessive compulsive disorder (OCD)	a pattern of repeated thoughts and behaviors that interferes with a person's life
phobia	an unreasonable fear of an object or event
paranoia	an unreasonable fear that someone is trying to harm you
psychosis	a serious mental disorder that must be treated by a doctor
psychiatrist	a medical doctor who specializes in treating serious types of mental disorders
schizophrenia	a kind of psychosis
depression	a state of mind that causes a person to withdraw and feel sad for long periods of time
anorexia	a disorder that causes a person not to eat
bulimia	a disorder that causes a person to overeat and then try to get rid of the food just eaten

Mental Disorder: What Does It Mean?

A **mental disorder** is a condition that disturbs a person's thinking and emotions. It can make a person unable to live comfortably with him or herself and others. A mental disorder can affect a person's behavior. He or she may stay away from people or act in unusual ways.

Neurosis

Mental disorders have many causes and forms. One group of mental disorders is called **neurosis**. A neurosis is any mild type of mental disorder that may be treated with therapy. A person with a neurosis is called a *neurotic*.

The most common neurosis is severe worry or anxiety. When **anxiety** attacks, the person has a feeling that something terrible will happen. He or she usually does not know what is causing the anxiety. The cause may be buried deep in the person's mind. It can cause tiredness, deep fear, sleeplessness, upset stomach, rapid breathing, and sweating.

People who feel the need to wash their hands every few minutes may have OCD.

Obsessive Compulsive Disorder

Obsessive compulsive disorder, or OCD, is a kind of neurosis that is caused when a pattern of related thoughts and behaviors interferes with a person's life. To be *obsessive* means thoughts repeat over and over again. For example, a person may think often about the germs all around him or her. These unwanted thoughts may prevent a person from thinking about normal things. To be compulsive means you want to do something over and over again. A compulsive person may wash his or her hands every 20 minutes to remove unseen germs. Often, people with OCD know their feelings and behaviors are not normal. But, they cannot seem to do anything about it.

There is help for people who have OCD. If thoughts or behaviors interfere with someone's daily life, he or she should seek help. He or she can talk with a doctor, parent or guardian, or a counselor about the problem.

Phobias, Paranoia, and Panic Disorders

Another kind of neurosis, called a **phobia**, is an unreasonable fear of an object or event. It can also cause strong anxiety. A person can have a phobia of just about anything. For example, people have phobias about dogs, cats, spiders, thunderstorms, dirt, and open or closed spaces. Some people have phobias of water, high places, or flying in airplanes.

Some people have an unreasonable fear that someone is trying to harm them. This neurosis is called **paranoia**. Extreme forms of paranoia can become very serious. People may do strange things to try to avoid an unknown danger.

Panic disorder is another form of neurosis. It is a fear or anxiety that takes over a person's life. The person may have panic attacks in which he or she shakes, has a faster heart rate, and feels short of breath. These panic attacks can be serious or mild. They may be caused by a variety of factors.

 **Health Fact**

Ailurophobia is a fear of cats. Claustrophobia is a fear of small places. Astraphobia is a fear of lightning. Hydrophobia is a fear of water. Microphobia is a fear of germs.

Many people have a fear of flying.

Help for Neurosis

Most people with a neurosis are able to lead normal lives in spite of their problems. They have jobs, friends, and families. But, if a neurotic person's life is affected by the neurosis, he or she may need help.

Psychologists are caregivers trained to provide therapy to people with mild mental disorders. They can help a person find the underlying cause of his or her neurosis. Then they can suggest ways to deal with it. Psychologists are not medical doctors, so they cannot prescribe medication.

Psychosis

The most serious mental disorder is called a **psychosis**. A psychosis must by treated by a doctor. A *psychotic* person relates less and less with others. He or she loses touch with the real world. Sometimes, the person does not react to things or people around him or her. Sometimes, the person may react to everything by showing very strong emotions.

Unlike neurotic people, psychotic people may be violent and dangerous. They need medical care. **Psychiatrists** are medical doctors who specialize in treating serious types of mental disorders. They can prescribe medicine for treatment.

The most common kind of psychosis is called **schizophrenia**. This illness has many symptoms, and no one knows exactly what causes it. A schizophrenic may act dull and uninterested for a while. Then, he or she may become wildly excited. He or she may smile and laugh while talking about some sad event. Sometimes, this person may not move or speak for days. He or she may be paranoid. Speech and thought may be very confused. This person is often in a world of sounds and sights that no one else can see or hear. He or she needs medical treatment.

Write About It

The term *schizophrenia* means "split mind." Why do you think this is an appropriate name for this disorder? Write your answer on a separate sheet of paper.

Other Types of Psychosis

There are many other types of disorders that are considered forms of psychosis. Each of these disorders has different causes. They also affect each person differently. Some of these disorders are worse than others. Below is a table that gives more information about other mental disorders.

Mental Disorders	
Disorder	**Description**
Hypochondria	fear of diseases that a person does not really have; focused on health; becomes anxious over minor aches and pains
Bipolar Disorder	manic depression; may be inherited; moods change rapidly between periods of extreme happiness and extreme depression
Antisocial Personality Disorder	behaving in an antisocial way; may do things that are dangerous to themselves or others; may do things that are illegal
Passive Aggressive Personality Disorder	being uncooperative with others; may have trouble following rules but does not confront authority directly; may be angry but does not deal with the issue causing the anger directly

✔ **Check Your Understanding**

On a separate sheet of paper, write the one word that each phrase describes.

1. a caregiver trained to work with mild mental disorders

2. a feeling that something terrible will happen

3. a fear that someone is trying to do you harm

4. a very strong fear of something

5. the most serious mental disorder

Depression

Talk About It

Why do you think so many young people suffer from depression?

Depression is a state of mind that causes a person to withdraw and feel sad for long periods. It often causes a person to feel like giving up. People suffering from depression do not like themselves or others. They may blame themselves for their misery. To depressed people, the world is a sad place with little hope.

Everybody has good moods and bad moods. But, when unhappiness does not go away, it becomes a cause for concern. Rather than being a temporary state of mind, depression can become an illness.

It is important for young people to know about depression. Studies show that as many as one in five teenagers suffers from depression. About 20 percent of people getting help for depression are under the age of 18. Depression is the leading cause of suicide. Suicide is a leading cause of death, after accidents, among American teenagers.

Signs of Depression

Depression can affect anyone. Here are some warning signs:

- eating less or eating more
- losing interest in activities
- inability to think clearly
- constantly tired; sleeping a lot
- feeling worthless or having low self-esteem
- thinking about suicide

Health Fact

Some people suffer from a kind of depression called Seasonal Affective Disorder (SAD). They get depressed in the winter when the days are shorter. Sitting in front of artificial sunlight for a few hours each day stops the depression.

Causes and Cures of Depression

Depression can be caused by crisis situations. Perhaps there is a divorce or a death in the family. Other things, such as failing in school, can also cause depression. Depression can be caused by a poor self-image or by feeling unloved. It can also be caused by a chemical imbalance in the brain. This chemical imbalance may occur naturally or it may be brought about by drug or alcohol abuse.

Sometimes talking to an adult, exercising, or eating healthier can cure a person's depression. Some types of depression can be cured by talking to a psychologist. Other types of depression must be treated with drugs prescribed by a psychiatrist. These drugs may not cure the depression. But, they may control it until a cause and cure are found.

Warning Signs of Suicide

It can be difficult to recognize that someone is depressed and thinking of suicide. People sometimes blame themselves for not helping a friend who has committed suicide. It is not unusual to feel this way. It is hard to know when someone is just feeling bad or feeling suicidal. The list below shows some clues that you can look for.

- A person seems depressed most of the time.
- A person says things are "hopeless."
- A person suddenly begins to give away things that are important to him or her, or makes a will.
- A person tells a friend or family member that he or she is thinking of suicide. In fact, 75 percent of people who try suicide tell someone about it first. Always take people seriously when they talk about suicide.

Talk About It

Almost 50 percent of young people who commit suicide are drunk or on drugs. Why do you think this is so?

- If you think a friend might be suicidal, do not try to solve the problem by yourself. Tell a family member or counselor about your friend. Keeping signs of suicide a secret may only end in a tragedy.

- Try to get your friend to talk about his or her feelings. Never make fun of or be angry with the person. Your friend needs help and support. Make sure he or she knows that you really care.

- Make sure someone is with the suicidal person all the time.

- Try to get your friend to see a mental health professional.

People in Health

SUICIDE HOTLINE VOLUNTEER

Luis Gonzales is a suicide hotline counselor. He is a volunteer. This means he does his job without being paid. Luis answers telephone calls from teens and adults who are depressed.

One of the most important things Luis does as a counselor is listen. Listening allows people who call the hotline to talk about their feelings and depression. Luis knows that if a person who has suicidal feelings or is depressed has called the hotline, he or she is looking for help.

Luis is an option for teens who do not feel comfortable talking to others about their feelings. He helps his community by trying to prevent suicides and improving public health.

CRITICAL THINKING Why do you think some teens might not feel comfortable talking to others about their feelings of depression and suicide?

Suicide hotlines are often run by volunteers.

Eating Disorders

Have you ever seen pictures of movie stars or Miss America winners from 50 years ago? Many of those in the photos look quite a bit heavier than movie stars and beauty pageant winners today. In fact, if you look at paintings of women from a few hundred years ago, the women will look much heavier than movie stars of today.

What makes a beautiful body? In societies where food is scarce, fat is considered very beautiful. As you know, in our society today thin is considered beautiful. Since the 1940s, a beautiful body is considered one that is quite thin.

People have all different kinds of body shapes and sizes. Unfortunately, many people do unhealthy things to their bodies to try to look like people on television, in movies, or in magazines. To become thin, young people often go on crash diets. Sometimes these diets turn into eating disorders.

Anorexia

People with **anorexia** are afraid to eat. They may be hungry, but the thought of food makes them feel sick. Usually, these people are already quite thin, but they see themselves as being overweight. These people lose 25 percent or more of their body weight. Anorexics are actually starving themselves to death.

Bulimia

People with **bulimia** eat in binges. This means they eat large amounts of food at one time. Then, they purge the food. This means they purposely try to get rid of the food. Sometimes, they take laxatives. Sometimes, they make themselves vomit right after eating. Bulimia can cause serious problems with the entire digestive system.

Talk About It

Where do you think society gets its ideas of what makes a body perfect?

Anorexia and bulimia are not just about losing weight. They are signs of deep emotional problems. People with these disorders do not have good control of their lives. They often have poor self-esteem. They are often depressed. People are especially prone to have these disorders in their teenage years. This is because they are going through so many life changes at once.

People with eating disorders are putting their lives and health in great danger. Poor nutrition can cause disease and death. People who think they have eating disorders should get professional help. A family doctor or a mental health professional is the place to start.

Talk About It

Some doctors say that anorexics are rebelling against their developing bodies. What do you think?

Eating disorders are very dangerous to a person's health. It is important that the person seeks professional help with his or her problems.

✓ **Check Your Understanding**

Write your answers using complete sentences.

1. What are three signs of depression?

2. What is the difference between anorexia and bulimia?

LIFE SKILL
Raising Awareness About Teen Suicide

Thousands of teenagers commit suicide each year. In fact, suicide was the third leading cause of death among 15- to 24-year-olds in 2000. Depression is one cause of suicide. Depressed people may feel helpless and hopeless. They may feel there is no way to change the things that make them feel bad. Suicide may seem like the only solution. Suicide may also be the result of extreme stress or anger. The use of alcohol and other drugs is often a factor in suicides.

WARNING SIGNS OF SUICIDE

- talks about death or committing suicide
- gives away some favorite possessions
- withdraws from activities that he or she previously enjoyed
- talks about getting even with someone who has hurt or made him or her angry
- ignores his or her personal appearance
- acts very differently
- gets lower grades in school
- experiences changes in eating habits
- participates in more high-risk behaviors
- has made previous suicide attempts

The key for preventing suicide in teenagers that show these signs is to get help before it is too late. Many communities have suicide hotlines or other crisis centers to help potential suicide victims.

Use the phonebook or the Internet to research a list of hotlines or centers in your community. Then, write each organizations name, address, and phone number on an index card.

Applying the Skill

Work with a partner to design a poster that identifies the warning signs of suicide. Make the poster informative and well designed. Include people to contact if someone suspects another person is suicidal.

Summary

Mental disorders are conditions that disturb a person's thoughts and behavior.

One group of mental disorders is called neurosis. People with neuroses can usually lead normal lives. Examples include anxiety, phobias, and OCD. A neurosis can be treated by a psychologist.

Psychosis is a more serious kind of mental disorder. Schizophrenia is one kind of psychosis. A psychosis must be treated by a psychiatrist or medical doctor.

Depression can be a serious mental disorder. Signs of depression that last more than a few days are cause for concern. Depression is a leading cause of suicide.

If you suspect a person is suicidal, it is important to get help for the person. If you recognize the warning signs of suicide, take the person very seriously.

anorexia

bulimia

depression

mental disorder

neurosis

psychosis

Vocabulary Review

Match each term to its definition. Write your answers on a separate sheet of paper.

1. a condition that disturbs a person's emotions, thinking, and behavior _____

2. a mild type of mental disorder that may be treated with therapy _____

3. a state of mind that causes a person to withdraw and feel sad for long periods _____

4. a disorder that causes a person not to eat

_____

5. a disorder that causes a person to overeat and then try to get rid of the food he or she ate

_____

6. a serious mental disorder that must be treated by a doctor _____

Chapter Quiz

Write your answers on a separate sheet of paper. Use complete sentences.

1. Name two main kinds of mental disorders. Give at least one example of each kind.

2. A person washes his or her hands 16 times a day for no reason. What might be wrong with the person?

3. What is a phobia? Give one example.

4. Describe how a schizophrenic person might act.

5. List six common warning signs of depression.

6. How might drug abuse be linked with depression?

7. What are three warning signs that someone is thinking about suicide?

8. What should you do if a friend tells you he wants to kill himself?

CRITICAL THINKING

9. Compare eating disorders and depression. How are they related?

10. What should you do if you think someone you know has a mental disorder that they do not realize?

Online Health Project

Depression can be a very serious illness. It can also be common in teenagers. Find out more about symptoms and examples of depression. How does it affect people your age? How can it be treated? Use your information to create a short television or radio public health announcement about teens and depression. Include information on resources people with depression can go to for support.

HEALTH LINKS℠
Go to www.scilinks.org/health.
Enter the code **PMH320** to research **depression**.

The pressures of daily life can make people feel stress. Exercise can be a good way to relax.

Learning Objectives

- Define stress and explain how it affects the body and mind.

- Identify major causes of stress.

- Describe ways to cope with stress.

- **LIFE SKILL:** Identify stress management techniques that fit your personality.

Chapter 14 ▷ Coping With Stress

Words to Know

stress	emotional pressure people feel when they face a difficulty
stressor	an event that causes stress
cope	to face or deal with a problem or responsibility
adrenaline	a hormone that gives the body extra energy and strength
stress response	an automatic reaction to a feeling of stress
visualizations	positive thinking used to control stress

What Stress Does to the Body

What happens to the body when you experiences stress? **Stress** is the emotional pressure people feel when they face a difficulty. Suppose you are playing in a big game. Your mind is racing with thoughts of what might happen.

The game is a **stressor**. It is an event that causes stress. Your nerves sense this stress and send a message to your brain. Your brain, in turn, sends a signal to your glands. Your glands then send out hormones that help you cope. To **cope** is to deal with the problem or responsibility.

One set of glands sends out adrenaline. **Adrenaline** is a hormone that gives the body extra energy and strength known as a stress response. A **stress response** is an automatic reaction to a feeling of stress.

These hormones can also help the body fight diseases that can be caused by the stress. But this effect does not last forever. A person who feels stressed day after day will soon become worn out. The body may lose its ability to fight off disease. Then, the person gets sick.

Talk About It

Think of the last event in your life that caused you stress. How did you feel, and how did you cope?

Stress can cause many illnesses. Some of these illnesses are heart disease and heart attacks, cancer, high blood pressure, depression, stomachaches and headaches, and muscle aches. Some doctors say that 75 percent or more of all human illness can be related to stress.

Recognizing Stress

Write About It

Have you ever had a physical illness that you think was caused by stress?

Sometimes people can go for weeks—even years—without knowing they are feeling stress. The first step in coping with stress is recognizing it. You can do this in two ways:

- by thinking about your feelings
- by looking at situations in your life

Have you ever had a bad day? Maybe you hit your brother, hung up on a friend, and yelled at your mother. Later you thought about what made you do those things. You had to admit that the problem was not really theirs—it was yours. Something was causing you stress, and you were angry.

Stressful Feelings

- lonely
- angry
- hurt
- afraid
- worthless
- excited
- depressed
- sad

It is important to think about your feelings. They hold important clues to what is causing you stress. Next time you are feeling depressed or upset, take a minute to describe what you are feeling. The list to the left shows some words that describe "stressful" feelings.

Once you describe these feelings, you can get in touch with what is causing them and your stress. Then, you can begin to cope with the problem.

You can also learn to recognize situations that cause you stress. By recognizing these situations, you can begin to understand where some of your feelings are coming from. Then, you can take steps to control stress before it happens. Stress can be caused by both positive and negative events in your life.

There are five main kinds of situations that cause stress:

Taking an important exam is an expected event that can cause stress.

1. **Major life events that you expect:** The first day of high school, marriage, and getting a new job are such events. A person may spend many stressful hours wondering what such events will bring.

2. **Unexpected events:** Deaths in the family, an accident, or becoming the victim of a crime are some examples.

3. **Situations that build up stressful feelings over time:** Daily pressures of school, work, or home can add up. Often, a person does not know that he or she is under stress. Then, one day, the person gets very angry. The stress has built up over time, little by little.

4. **Feelings about yourself:** How you feel about yourself can cause stress. If you have a poor self-image, almost any situation will cause you stress. If you have a fear of failure, taking risks and trying new things will cause stress.

5. **Situations that test your values:** Suppose you must choose between using drugs or losing your friends. Your friends are important to you, but your health is also very important to you. This kind of conflict can cause great stress in a person.

✓ Check Your Understanding

Write your answers in complete sentences.

Talk About It

Some people feel stressed during holidays. Why do you think this happens?

1. Write down a major event from your life. Did it cause you stress? If yes, explain how you felt. If no, explain why you were not stressed.

2. Write down a situation that tested your values. Did it cause you stress? If yes, explain how you felt. If no, explain why you were not stressed.

Think About It

DOES ALTERNATIVE MEDICINE REALLY WORK?

During the past 20 years, *acupuncture* has become a popular type of alternative medicine. In this process, a needle is placed through the skin of the sick person. The location of the needle varies according to the illness. That is because acupuncturists believe that different locations, or points, control different body parts.

Americans spend as much as $500 million a year on acupuncture. Scientists have studied the process. They have found that the points are connected to the central nervous system. The researchers believe that placing needles through these points releases chemicals that reduce pain and stress. One study showed that acupuncture lessened back pain in 50 percent of patients and headache pain in 70 percent of patients. Many people think acupuncture can also relieve stress.

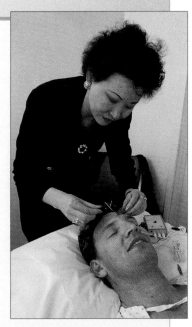

Many people believe acupuncture can relieve stress.

Hypnotherapy is another form of alternative medicine used to treat pain and stress. In this process, a hypnotist focuses the patient's attention on the injured body part. The patient concentrates on how the body part would look if it was healthy. It is believed that forming this mental image helps relieve pain. While many people agree, more research is needed in this area to determine the actual medical benefits.

YOU DECIDE Do you believe acupuncture or hypnotherapy can really ease pain or relieve stress? Explain why or why not.

How to Cope With Stress

Too much stress can hurt your health. People who are always feeling stressed are often angry or sad. If you find you are angry and sad a lot of the time, you may want to find ways to control your stress. Here are some ways to do that:

- Spend 10 minutes each day by yourself. Turn on some soft music, or sit in a quiet room. Close your eyes and think about positive things or nothing at all. Take deep breaths and relax. This will calm you and help you to prepare for stressful situations.

- Fast-paced exercise will get rid of stress and make you feel better about yourself.

- **Visualizations** can help you control stress. Visualizations are a kind of positive thinking. For example, suppose you are about to take a driver's test. Do not think about how hard it will be. Instead, picture yourself doing a great job on it. See the person testing you saying how well you drive. Picture yourself putting your new driver's license in your wallet.

- Try to become a problem solver. If you get into a car accident or face another crisis, say to yourself: "Here is a problem. How do I solve it?" If you calmly take control of the situation, your stress will lessen.

- If you are often stressed, talk about your feelings to someone you trust. Remember that recognizing your feelings, and what may be causing them, is the first step toward becoming a problem solver.

Write About It

How do people in your family cope with stress? You may have learned to manage your emotions the same way they do. Are there any changes that need to be made? Write your answer on a separate sheet of paper.

Health & Safety Tip

Studies have shown that simply making yourself smile will make you feel better.

Write your answers in complete sentences.

1. List four ways you can relieve stress.

2. How can visualizations help you control stress?

3. What do you think it means to "become a problem solver?"

4. CRITICAL THINKING Your family is planning to move to another state. They want you to help them plan and pack. How will you cope with the stress and make sure your family gets help?

Think About It

ARE SLEEPING PILLS A GOOD IDEA?

People who are feeling a lot of stress sometimes have trouble sleeping. They stay awake worrying about things, or they do not sleep well.

Sleeping pills are drugs prescribed by doctors. Some people say that the pills help break the cycle of stress and difficulty sleeping. A person who has not been able to sleep is going to stay stressed. But, if that person can sleep with the help of a sleeping pill, then the stress will go away. Also, the need for the sleeping pill will go away.

Other people say that taking prescription drugs is a bad way to solve a problem. They say that the stressed person should try to look at the feelings and conditions causing the stress. The pill will only make the person sleep. It will not solve the problem causing the stress.

YOU DECIDE Do you think taking sleeping pills is a good way to handle stress or deal with problems? Why or why not?

Sleeping pills may help people to relieve stress, but they do not always solve the problem that is causing the stress.

LIFE SKILL
Practicing Stress Management Techniques

Everyone experiences some stress. A little stress is good. It can make you feel excited and energetic. But, too much stress can be harmful. Too much stress can lead to health problems.

Adolescence can be a very stressful time. It is important to know how to manage the stress in your life. Here are some healthy ways to manage stress:

A hobby can help you manage stress. Doing things you enjoy can relax you.

- Find ways to feel good about yourself. You can learn a new skill or take up a new hobby. Doing volunteer work is also a good way.

- Take time to relax. Spend 10 minutes each day by yourself. Close your eyes and think about positive things or nothing at all.

- Try to plan your time well. Leave enough time to do things. Then, you will not feel hurried. Plan to get important tasks done first.

- Find good friends who accept you as you are. This will allow you to relax and be yourself.

- Learn how to express your feelings and settle arguments peacefully.

- Eat a balanced diet and exercise.

- Ask for help. If you cannot handle everything you have to do, find someone to help you.

Suppose that you feel stress because of a change in your family. Someone tells you to accept the things you cannot change. Do you think that is good advice? Explain why or why not. Give examples of other advice that someone could give.

Applying the Skill

List some causes of stress in your life. Write a paragraph telling how you might manage that stress.

Summary

Stress is the emotional pressure people feel when they face a difficulty. When the body experiences stress, glands make special hormones. These hormones give a person extra energy and strength. They help the body fight disease. If stress is experienced over a long time, the body tends to lose its ability to fight disease and gets sick.

Stress can be caused by different expected and unexpected events. Stress can also come from building up bad feeling about the things around you or yourself. Situations that test your values can cause stress.

To cope with stress, you must learn to recognize stressful feelings and events. You must learn ways to control stress. One of the most important ways is to become a problem solver.

adrenaline

cope

stress

stressor

stress response

visualizations

Vocabulary Practice

Match each term to its definition. Write your answers on a separate sheet of paper.

1. positive thinking used to relieve stress

_____

2. an event that causes stress _____

3. emotional pressure people feel when they face a difficulty _____

4. to face or deal with a problem or responsibility

_____

5. a hormone that gives the body extra energy and strength _____

6. an automatic reaction to a feeling of stress

_____

Chapter Quiz

Write your answers on a separate sheet of paper. Use complete sentences.

1. What is stress?

2. Give an example of a stressor.

3. What does it mean to cope with something?

4. What does the body do when it experiences stress?

5. Why do some people who lead stressful lives get sick?

6. Give three examples of stressful feelings.

7. How can a happy event such as a wedding cause stress?

8. What situations might cause stress?

9. What do your personal values have to do with stress?

10. Name two ways you can control stress.

CRITICAL THINKING

11. Which types of stress management techniques listed on page 207 do you think would work best for you?

12. Explain how stress affects physical health.

Online Health Project

Relieving stress has become big business in the United States. Companies can make a great deal of money selling products that are supposed to ease stress and make your life better. These products may or may not actually work. Research several products that claim to relieve stress. Find out whether or not these products have any medical or scientific data to support their claims. Then, write a report that analyzes the validity of these products.

HEALTH LINKS℠

Go to www.scilinks.org/health. Enter the code **PMH330** to research **stress**.

Unit 3 **Review**

Comprehension Check

On a separate sheet of paper, explain how each of the following items prevents disease or keeps your mind and body healthy.

1. Talking about your feelings

2. Exercising when you are angry

3. Doing activities that help you feel positive about yourself

4. Having a positive attitude

Analyzing Cause and Effect

Write a sentence or two explaining the possible causes of the following behaviors. The information you learned in Chapters 11–13 will help you.

5. A parent abuses his or her children.

6. A father and son are both "nice guys."

7. A woman is so afraid of cats that she screams when she sees one.

8. A person refuses to take heart medicine. He or she says, "Nothing bad could ever happen to me."

9. An athlete spends at least 10 minutes a day picturing him or herself winning a race in record time.

Writing an Essay

Answer the questions below on a separate sheet of paper. Use complete sentences.

10. Why do people need defense mechanisms?

11. Explain how stress affects your health.

12. How could running 2 miles a day be good for your emotional health?

13. Name at least two health professionals who specialize in treating mental illness and emotional problems.

14. Why is it so important for teenagers to know about depression?

Managing Your Health

On a separate sheet of paper, list five things that could cause you stress. Then, list five ways you could cope with that stress in healthy way.

Before You Read

In this unit, you will learn about legal and illegal drugs. You will learn about the health risks of smoking and alcohol. You will also learn how illegal drugs affect a person's emotional health.

Before you read, ask yourself the following questions:

1. What do I already know about the health risks of smoking?

2. What questions do I have about alcoholism?

3. Are legal drugs and medicines completely safe?

These students are gathered at a smoke-out event. Events like these are organized to promote a healthier lifestyle.

Learning Objectives

- Define what a drug is.

- Define medicine and give some examples.

- Name some common reasons why people use alcohol and tobacco.

- Describe how alcohol and tobacco affect the body.

- Explain the difference between a stimulant and a depressant.

- LIFE SKILL: Analyze the influence of the media on decisions about smoking.

Chapter 15 ▷ Medicines and Drugs

Words to Know

drug	a chemical substance that affects the body systems
medicine	a drug that is used to prevent or cure a disease or medical problem
over the counter medicine (OTC)	medicine that can be bought without a doctor's prescription
prescription medicine	medicine that must be ordered by a doctor
immunization record	a record that shows which vaccines a person has received
depressant	a drug that slows down the nervous system
blood alcohol level (BAL)	a measurement of how much alcohol is in the blood
alcoholism	a disease that causes a person to be addicted to alcohol
alcoholic	a person who is addicted to alcohol
cirrhosis of the liver	an often fatal disease that causes the liver to stop functioning
addiction	a need or habit for a substance
nicotine	the addictive drug in tobacco
stimulant	a drug that speeds up the nervous system and other body systems
withdrawal	a physical reaction to the removal of an addictive substance in the body

What Are Drugs?

A **drug** is a chemical substance that affects your body systems. Aspirin is a drug. Alcohol and tobacco are also drugs. People do not usually think of these types of drugs as very harmful. After all, if they were harmful, why would they be legal? But, all drugs affect the body and can be dangerous. It is important to know as much as you can about these drugs. Then, you can make a wise decision about whether or not to use them.

Medicines

Medicines are drugs that are used to prevent or cure diseases or medical problems. Antibiotics are examples of medicines. There are two main types of medicines. Medicines called **over the counter medicines**, or OTCs, can be bought without a doctor's prescription. **Prescription medicines** must be ordered by a doctor.

Groups of Medicines

Medicines can be grouped by their effect on the body. Examples of the different types of medicines are medicines that prevent diseases, fight pathogens, relieve pain, and help the circulatory system.

A *vaccine* is an example of a medicine that prevents diseases. The vaccines that you are given depend on your age. You probably were given a lot of vaccines when you were younger. A record was probably kept that shows which vaccines you have received. This record is called an **immunization record**.

Antibiotics are a type of medicine used to kill pathogens. Many antibiotics have been developed. Each antibiotic works against a specific group of pathogens. You may have taken antibiotics if you had strep throat, or a sinus infection. Antibiotics can come in liquid or pill form.

A wide variety of medicines are available today. It is important to know what each one is used for.

Health & **S**afety **T**ip

When a doctor prescribes an antibiotic for you, it is important that you finish all of the medicine. This is to prevent the pathogens from becoming resistant to the antibiotic.

Other medicines are used to help the circulatory system work properly. These medicines may regulate heart beat, lower blood pressure, or help keep veins and arteries open.

Using Medicine Safely

It is important for your health to use medicines safely. You should know the side effects of a medicine and how it is supposed to be taken.

A *side effect* is a reaction to medicine that is different from the reaction the medicine is supposed to cause. For example, some antibiotics may cause an upset stomach. The upset stomach is a side effect. You can find out what side effects are possible with a certain medicine by reading the label or asking your healthcare provider.

It is also important to use medicines in the way they are intended. Below is a list of guidelines for how you can use medicines safely.

Health Fact

There are many pain relievers available today. The pain reliever that is right for you depends on your age, body weight, and the health problem that you have.

USING MEDICINES SAFELY

- Always read the directions before taking any medicine.
- Always use the correct amount of medicine.
- Never use someone else's medicine. Also, never give your medicine to someone else.
- Never mix medicines without talking to your doctor first.

✓ Check Your Understanding

Write your answers in complete sentences.

1. What is a drug?

2. What are medicines? Give some examples.

3. What are three ways you can use medicines safely?

Alcohol and the Body

Alcohol comes in many forms: beer, wine, and hard liquors, such as vodka, rum, or whiskey. Once swallowed, alcohol goes quickly into the blood. The blood carries the alcohol to the brain. There, the alcohol acts as a depressant. **Depressants** are drugs that slow the nervous system. Alcohol is also a *psychoactive drug*. It changes the way a person's brain functions.

Many people feel alcohol helps them to relax. But, alcohol has many harmful effects on the body. Below is a description of short-term and long-term effects of alcohol use.

Alcohol comes in many forms.

Short-Term Effects of Alcohol Use

- impaired judgment
- slower reaction time and reflexes
- slurred speech
- brain cells are killed
- unconsciousness
- poor vision and hearing
- loss of muscle control and balance
- body becomes dehydrated
- vomiting
- death, if used in large amounts

Long-Term Effects of Alcohol Use

- permanent liver damage, or cirrhosis of the liver
- damage to the heart and stomach
- causes emotional and social problems for the alcohol user and others
- leads to cancer of the liver
- damage the user's ability to learn and remember

Talk About It

60 percent of murders, 33 percent of suicides, and 65 percent of family violence are alcohol related. How can you explain these statistics?

Alcohol also affects a person's mood and personality. Some people become very friendly and talk a lot. Others become sad and depressed. Some people want to fight and argue.

Blood Alcohol Level

Suppose you are driving, and a police officer stops your car. If the officer suspects that you have been drinking, you will have to take a test. The test will measure the amount of alcohol in your blood, or your **blood alcohol level (BAL)**.

An average person's body can break down one ounce of alcohol per hour. If more than one drink goes into the body per hour, the blood alcohol level rises. The person becomes drunk. In most states, a person is legally drunk when his or her blood alcohol level reaches 0.1 percent. The chart on the right shows what happens at the different blood alcohol levels.

✓ **Check Your Understanding**

Write your answers in complete sentences.

1. Why is alcohol considered a psychoactive drug?

2. List four effects of alcohol use.

Alcohol Level In Blood	
(%)	Effects
0.05%	careless behavior; some loss of self-control; some loss of coordination
0.10%	poor judgment; greater loss of self-control; serious loss of coordination
0.20%	very drunk; slurred speech; staggers when walking
0.40%	unconscious; passes out
0.70%	death occurs

Think About It

DO BREATH ANALYZERS SAVE LIVES?
Drinking alcohol can cause a person to drive very dangerously. It is important for police officers to be able to test a driver to find out how high his or her BAL is. A tool called a Breath Analyzers is used to measure the amount of alcohol in a driver's breath.

How do Breath Analyzers work? Alcohol is not digested. Instead, it is absorbed into the bloodstream. Alcohol in the blood that flows through the lungs, leaves the body when air is exhaled. The amount of alcohol in the breath depends on the amount of alcohol in the blood. A Breath Analyzer can be used to find the BAL amount in a person's blood.

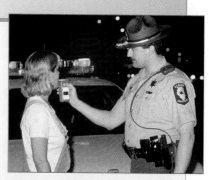

This police officer is using a Breath Analyzer.

YOU DECIDE How do Breath Analyzers keep people safe?

Alcohol and Motor Vehicle Safety

Health Fact

About 50 percent of all motor vehicle accidents are caused by people who have been drinking alcohol.

Alcohol begins to affect the body very quickly. Even one drink can have a negative effect on a person's ability to drive a car. Think about the short-term effects of alcohol listed on page 216. Impaired judgment, poor vision, and slower reaction time are just a few of the effects of alcohol. Do you think a person can safely drive a car under these conditions?

When a person's blood alcohol level has risen to about 0.08 percent–0.1 percent, he or she is considered *intoxicated* in many states. People who choose to drive while intoxicated are breaking the law. This crime is called *driving while intoxicated* (DWI), also called *driving under the influence* (DUI). Driving while intoxicated or driving under the influence is a serious crime. It can result in the loss of the person's driver's license, fines, probation, and time in jail. Although a person is considered legally drunk with a blood alcohol level of 0.1 percent, many states are changing this law. It has been found that a person actually starts to lose driving skills at around half that level, or 0.05 percent.

Health & Safety Tip

To protect your health, never get into a car with someone who has been drinking.

Driving after drinking alcohol has even more harmful consequences than time spent in jail. Car accidents involving alcohol are often very serious. Many of these accidents are fatal. In fact, more than 17,000 Americans are killed each year in alcohol-related motor vehicle accidents.

Alcohol related car accidents can be very serious and even deadly.

Alcohol Addiction

Most adults who drink alcohol are called *social drinkers*. They have no more than a couple of drinks now and then, usually with friends. They do not drink to escape life's problems. They can tell when they have had enough to drink. They act responsibly.

However, there are people who do not know their limits. **Alcoholism** is a disease that causes people to become addicted to alcohol. People with alcoholism are called **alcoholics**. This means that both their bodies and minds crave more and more of the drug. Job, family, and social life all take second place to getting enough alcohol.

Years of drinking can damage some body parts beyond repair. For example, alcohol kills brain cells. As a result, the ability to learn and remember is damaged or destroyed. The liver, heart, and stomach can also be damaged. Alcohol can cause **cirrhosis of the liver**, a disease that stops the liver from working. Cirrhosis can kill. If a pregnant woman drinks, she may be addicting her baby to alcohol. An **addiction** is a need or habit for a substance. Also, unborn babies of women who drink are more likely to have health problems after birth.

Families of Alcoholics

Alcoholism is a disease that affects everyone around an alcoholic. Everyone in a family is hurt by the alcoholic's drinking. These people often make the disease worse for the alcoholic without knowing it. For example, they might hide the alcoholic's poor work or bad behavior.

Write About It

Alcoholics can get much of the energy they need from the calories in alcohol. But, alcohol has very few nutrients. How do you think this lack of nutrients affects the body?

Health Fact

Babies born to mothers who drink while pregnant sometimes have *fetal alcohol syndrome*. This disease causes a low birth weight, mental disabilities, and an abnormally shaped head and other body parts.

Children of alcoholics have a special set of problems. Their needs are often not met because of their parent's addiction to alcohol. Sometimes they try very hard to please other people. Children may feel responsible in some way for the alcoholic's disease.

Alcoholism and Teenagers

It is illegal for people under the age of 21 to drink alcohol in the United States. But, many teenagers choose to drink anyway. Some teenagers drink alcohol because they think it will help them fit in or be more popular. Some teenagers drink alcohol to feel more relaxed. Alcoholism is a growing problem among teenagers. In fact, it is very easy for a young person to become addicted to alcohol. A teenager can become an alcoholic in a matter of months.

Even though teenage drinking has become a growing problem, many teens today are choosing to stay away from alcohol. They understand the health risks that go along with drinking alcohol. The simplest way to avoid these health risks is to *abstain* from alcohol. To abstain means you choose not to drink any alcohol. Saying no to alcohol can be hard in some social situations. It is a good idea to practice what you are going to say before the situation arises. That way, you will feel comfortable with your answer. You will also feel confident that you made the right decision.

Are You or Is Someone You Know a Problem Drinker?

Read through the checklist on page 221. In your mind, check the sentences that are true for you. Think about someone you think might have a problem. A check in any area is a sign that he or she may have a drinking problem. If so, it would be a good idea for that person to get help now.

Talk About It

What are some things teenagers can do to have fun without drinking?

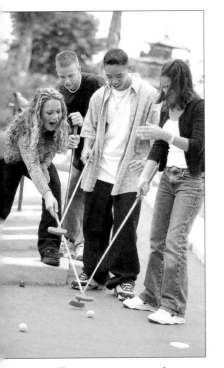

Teenagers can enjoy being with friends without drinking alcohol.

Read the signs of a drinking problem. Ask yourself, does the person:

- drink to forget or escape problems?
- have to drink to have a good time at a party?
- miss school or work because of drinking?
- drink alone?
- drink more than before?
- get into trouble because of drinking?
- do poorly on schoolwork or at work because of drinking?
- drink more than his or her friends do?

Where to Get Help

Many alcoholics quit drinking and recover from their disease. Once people recognize they have the disease, they can never drink again—even a little—if they want to recover. Most people need help to quit drinking.

Community mental health clinics and doctors are good places to get help. Alcoholics Anonymous (AA) is a group that has helped many alcoholics recover. Al-Anon helps families and friends of alcoholics. To find these groups near you, look in the yellow pages of a telephone book under *alcoholism*.

✓ Check Your Understanding

Write your answers in complete sentences.

1. At what blood alcohol level is driving affected?

2. What are some effects of alcoholism?

3. CRITICAL THINKING What are some ways alcoholism affects society?

Tobacco and the Body

Write About It

More than 3,000 teenagers become smokers every day. Why do you think teenagers start smoking when so much information is available about the dangers of smoking?

Cigarettes, cigars, pipes, snuff, and chewing tobacco all contain tobacco. Tobacco is harmful to your health. Tobacco has an addictive drug in it called **nicotine**. Nicotine is a **stimulant**. It causes the nervous system to speed up. The digestive and circulatory systems also work faster when nicotine enters the body. Nicotine causes the heart to beat faster and makes the blood vessels more narrow. Eventually these blood vessels can become so narrow that the smoker has a heart attack or stroke.

Tobacco also contains harmful tars and chemicals. These substances collect in the lungs and form a thick, brown, sticky substance. This material prevents the lungs from working properly, causing lung disease. It has been proven that smoking causes lung cancer. Lung cancer is currently the most common form of cancer in the United States. It is a deadly disease, but you can lower your risk of getting lung cancer by not smoking. Smoking also leads to a disease called emphysema. *Emphysema* is a serious disease of the respiratory system. It causes a person to have difficulty breathing.

 **Health Fact**

About 90 percent of men and 50 percent of women who die of lung cancer are smokers.

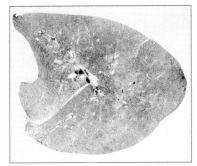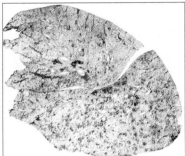

A nonsmoker's lung is pink and healthy (left). A smoker's lung is discolored and unhealthy (right).

Here are some more facts about how tobacco affects the body:

Remember
Cilia are hairlike structures that line the airways and lungs.

- Smoking destroys the cilia in the lungs. Without the cilia, dirt irritates the lungs. This can cause *bronchitis*, a coughing condition.

- People with chronic (repeated) bronchitis often get emphysema.

- Most smokers become sick more often than nonsmokers.

- Smokers have a higher risk of getting cancer of the lungs, mouth, and throat.

- Smokers have higher blood pressure than nonsmokers.

- Most smokers die earlier than nonsmokers.

- Babies born to mothers who smoke weigh less than babies of nonsmoking mothers.

Tobacco Addiction

Tobacco is a very addictive substance. People who are addicted to nicotine have trouble quitting even if they want to. They develop a habit. When smokers stop using tobacco, they go through a period of **withdrawal**. Withdrawal is a physical reaction to the lack of an addictive substance in the body. Withdrawal from nicotine can result in headaches, mood changes, and anxiety. Many smokers try to quit several times without succeeding. Some smokers use a nicotine patch or nicotine gum to help them quit.

Health & Safety Tip

The best way to avoid becoming addicted to nicotine is to never start smoking.

Even if you are already addicted, you can decide to quit at any time. There is help available to those who want to quit. Support groups can help someone who is struggling to quit. Doctors can also prescribe medication that can help a person quit smoking.

Tobacco and Nonsmokers

Nonsmokers who breathe in smoke from smokers suffer harmful health effects just as smokers do. When a person smokes, he or she exhales some of the smoke into the surrounding air. This smoke is often called *secondhand smoke*. Secondhand smoke can make a person's eyes water and become itchy. It can also cause headaches, coughing, and a sore throat. Recent studies have shown that secondhand smoke can also cause heart and lung disease if a person is exposed to the smoke for a long period of time.

✓ Check Your Understanding

Write your answers in complete sentences.

1. What is the drug in tobacco called?

2. What are three diseases related to tobacco?

3. CRITICAL THINKING Why is quitting smoking so difficult?

Think About It

HOW DO SMOKING LAWS AFFECT PEOPLE?
Many public places, such as restaurants and shopping malls, do not allow smoking. Other places allow smoking only in certain areas. Some states have made it illegal to smoke in all public places.

These laws usually make nonsmokers happy. They do not want to breathe smoke-filled air. But, some smokers do not like the laws. They feel they should be able to smoke where they want.

The laws are meant to protect nonsmokers. But, are they affecting the rights of smokers as well?

YOU DECIDE
1. Do you think these laws are fair? Why or why not?

2. Do these laws affect the rights of smokers?

THANK YOU FOR NOT SMOKING IN THIS AREA

PLEASE USE DESIGNATED SMOKING AREA

MALL OF ORANGE

Many public places do not allow smoking in certain areas.

LIFE SKILL
Analyzing Mixed Messages About Smoking

Sometimes you get mixed messages about smoking. Adults tell you that smoking is bad, even though some adults themselves smoke. Adults tell you it is addictive. They say it can kill you. Yet, movies, billboards, and magazines never show people with stained teeth. They do not show people coughing or with cancer. Ads and movies usually show smokers as attractive and healthy.

These images can be hard to resist. Many teens feel unsure of themselves. They want to fit in. Tobacco advertisements appeal to these feelings. They show smoking as a way to look good and be cool.

Can this affect your decision to smoke? Each day, thousands of people under age 18 start smoking. Most of them say they want to stop smoking, but they have a hard time quitting.

Tobacco ads usually show healthy, happy people.

Look at the advertisement. What is its message? Explain what it is that the advertisers are trying to make you to think. Write your answer on a separate sheet of paper. Use complete sentences.

Applying the Skill
Create an antismoking ad. Use images that will interest teens. Share your ad with the class.

Summary

A drug is a substance that affects your body and mind.

Medicines are drugs that are used to prevent or cure a disease or other medical problem. Medicines should be used with caution.

Depressants are drugs that slow down your nervous system. Stimulants speed up the nervous system.

Alcohol is a depressant. It causes a person to lose self-control and judgment. Drinking and driving is especially dangerous. Large amounts of alcohol damage both the body and brain. Alcohol can even cause death.

Alcoholics are addicted to alcohol. Alcoholism is a disease. It hurts everyone around the alcoholic as well.

Tobacco is a stimulant. Smokers are addicted to the nicotine in tobacco. Nicotine, tar, and other chemicals in the smoke can destroy the lungs. They can cause bronchitis, emphysema, and cancer of the lungs, mouth, and throat.

The decision to use alcohol and tobacco is an ongoing one. You can always decide to quit.

alcoholism

depressant

drug

medicine

nicotine

Vocabulary Review

Complete each sentence with a term from the list.

1. A drug that is used to prevent or cure a disease or medical problem is called _____.

2. A chemical substance that affects the body systems, brain, and behavior is called a _____.

3. A disease that causes a person to be addicted to alcohol is called _____.

4. The addictive drug in tobacco is called _____.

5. A drug that slows down the nervous system is called a _____.

Chapter Quiz

Write your answers on a separate sheet of paper. Use complete sentences.

1. In your own words, what is a drug?

2. What are the two main types of medicines?

3. How can medicines be used safely?

4. What kind of drug is alcohol?

5. What are three health risks of drinking alcohol?

6. What happens to families of alcoholics?

7. Where could you go if you or someone you knew had a problem with alcohol?

8. Compare a stimulant and depressant. How are they alike or different?

9. What is secondhand smoke?

10. List three health risks that smokers face.

CRITICAL THINKING

11. Suppose a friend tells you that drinking beer cannot hurt you. Is he or she right? Give your reasons.

12. Have you ever felt peer pressure to use alcohol or tobacco? Explain the risks of using either drug.

Online Health Project

Drunk driving kills thousands of people every year. You may even know someone who has been in an accident caused by a drunk driver. Research the ways communities and government agencies are working to prevent drunk driving. Find out what you can do to help eliminate this serious health problem. Create a plan for making a difference in your community.

HEALTH LINKS.
Go to www.scilinks.org/health.
Enter the code **PMH340** to research **drunk driving**.

People can have fun with friends without drugs.

Learning Objectives

- List some reasons why people abuse drugs.
- Describe the difference between drug use and drug abuse.
- List the four main types of drugs. Give an example of each type.
- Explain the effects of abusing certain drugs.
- Explain how drugs can affect a community.
- Describe where you can get help for a drug problem.
- **LIFE SKILL:** Practice refusal skills.

Words to Know

barbiturate	a drug that slows down the nervous system
addictive	habit forming
psychological dependence	a person's emotional need for a drug
tranquilizer	a drug that has a calming effect
amphetamine	a drug that stimulates or speeds up the nervous system and heart rate
narcotic	a drug that comes from the opium plant
hallucinogen	a drug that causes a person to see, hear, and feel things that are not real
Cannabis sativa	the scientific name for marijuana
inhalant	a substance that is inhaled

Drug Use and Drug Abuse

Many kinds of drugs are used as medicines. Taking a drug such as aspirin can help relieve a headache. But, suppose a person takes 12 aspirin pills just to see what happens. The person is no longer using aspirin. The person is abusing it.

People abuse legal and illegal drugs for many reasons. The most common reasons are to treat pain or illness, curiosity, peer pressure, to forget about troubles, or to have a good time. Drug abuse has many harmful, and even fatal, consequences.

Types of Drugs Illegally Abused

Different drugs affect the body in different ways. Alcohol is a depressant. It slows down body functions. Nicotine is a stimulant. It speeds up body functions. These drugs are legal. Some illegal drugs are also depressants or stimulants.

Two other types of illegal drugs are called narcotics and hallucinogens. The chart below gives a brief explanation of these four types of illegal drugs.

Types of Drugs Illegally Abused		
Type of Drug	**Examples**	**How It Affects the Body**
Depressant	barbiturates; tranquilizers	slows down the nervous system; causes confusion and impaired judgment
Stimulant	amphetamines; cocaine	speeds up heart rate and breathing; causes anxiety and trouble sleeping
Narcotic	heroine; morphine	causes sleepiness, nausea, and slower breathing rate
Hallucinogen	LSD; PCP	confuses the brain and sense organs; user sees or hears things that are not real

Remember

A depressant is a chemical that slows down body functions. A stimulant is a chemical that speeds up body functions.

✓ Check Your Understanding

Write your answers in complete sentences.

1. How is drug use different from drug abuse?

2. Name three examples of stimulants.

3. Which type of drug causes sleepiness and slow breathing?

4. CRITICAL THINKING A friend has some painkillers that were prescribed for a broken arm. He wants to let some friends take them for fun. Explain how this is a form of drug abuse.

Barbiturates

Barbiturates are depressant drugs. Like alcohol, they slow down the nervous system. Barbiturates are often called *downers*.

Doctors sometimes prescribe barbiturates for people who cannot sleep or relax. Barbiturates lower blood pressure by slowing down muscle action, breathing, and heart rate.

People who abuse barbiturates enjoy the sleepy, pleasant feeling the drug gives them. But, they may also feel confused and have slower reflexes. A large dose of barbiturates can cause kidney poisoning and death. When mixed with alcohol, barbiturates can cause a person to lose control of his or her body.

When a person dies of an accidental drug overdose, barbiturates are often to blame. The abuser forgets how many pills were swallowed and takes some more. The overdose causes death.

Barbiturates, like many drugs, are habit-forming or **addictive**. Over time, the user thinks he or she cannot get along without the drug. This need is called **psychological dependence**.

Tranquilizers

Tranquilizers are drugs that doctors may prescribe to help a person relax. They are a type of depressant. Unlike barbiturates, tranquilizers do not make the user sleepy. Like other drugs, tranquilizers are very addictive and can cause physical and psychological dependence.

The drug ketamine is a powerful animal tranquilizer. It is sold on the streets as *Special K*. It can cause a person to see things that are not really there. Special K can also cause a person to feel as if they are unable to move. Rohypnol is another type of tranquilizer drug. It can cause people to become unconscious or fall asleep.

Health Fact

Barbiturates come in many forms. Some common nicknames are "barbs," "black beauties," and "yellow jackets." If someone offers you one of these, say "no."

Amphetamines

Amphetamines are stimulants. Like tobacco, they stimulate or "speed up" the nervous system, brain, and heart rate. Amphetamines are often called *uppers*. Other common nicknames for amphetamines are *meth*, *crank*, *ice*, and *truck drivers*.

A person who abuses amphetamines feels excited, happy, and powerful for a while. But, when the drug wears off, the person feels tired and let down. This is because amphetamines force the body to use energy faster.

Amphetamines also take away a person's appetite. Sometimes people take amphetamines as diet pills to lose weight. This may cause the person to want another stimulant. This is how the cycle of addiction starts.

Like barbiturates, amphetamines can cause psychological dependence. A person addicted to amphetamines can get very sick. Loss of weight, sleeplessness, confused thinking, shakiness, and blurred vision are signs of amphetamine abuse.

One of the most powerful amphetamines is *Methedrine*, or "speed." It is a whitish powder made into pills or capsules. Some Methedrine abusers sniff the powder. Others mix it with water and inject it into their veins. When the blood carries the drug to the brain, the abuser gets a sudden, strong feeling called a *rush*.

Speed abusers often collapse after a few days of using the drug. If they do not use more of the drug, they become depressed. Sometimes they commit suicide if they cannot get more speed.

Amphetamines often come in different pill forms.

Methedrine can come in the form of white rocks.

Cocaine and Crack

Cocaine is also called *coke*. It is a stimulant. People often take cocaine as a powder through the nose. This is called snorting. Cocaine can also be injected or smoked.

Like amphetamines, cocaine speeds up the heart and blood pressure. People who abuse cocaine often feel happy and excited. They have great feelings of power. They believe that nothing can hurt them.

Cocaine is a very dangerous, addictive drug. In fact, a person can become addicted after using it only once. Once addicted, a cocaine abuser may become difficult to be around. A cocaine abuser finds that everything takes second place to the drug. Cocaine is a costly drug. Some abusers spend more than $1,000 per day for their drugs.

It used to be that the cost of cocaine kept many people from using it. Then, drug abusers started using crack instead. Crack is a form of cocaine that costs $5 to $25 per dose. Abusers who smoke crack get an immediate high. The high lasts 5 to 20 minutes. Then, the person becomes depressed. Smoking crack causes a great deal of stress on the body. Crack abusers of any age are risking heart attacks and death.

Crack cocaine is a very dangerous drug.

 **Health Fact**

Using cocaine damages the tissues of the nose and causes bleeding.

✓ Check Your Understanding

Write your answers in complete sentences.

1. What are four drugs that are often abused?

2. What does *psychological dependence* mean?

3. CRITICAL THINKING How has the selling of crack changed drug abuse?

Heroin

Talk About It

Many burglaries and robberies are committed by people trying to get money for crack or heroin. Explain this fact.

Heroin is a narcotic. A **narcotic** is a drug that comes from the opium plant. Narcotics slow down the body functions a lot. They can even cause a coma or death.

Heroin is also called *junk*, *horse*, and *smack*. Heroin is a white powder that is usually mixed with water and injected. A heroin abuser feels relaxed and sleepy until the drug wears off.

Without the drug, the heroin addict goes through terrible withdrawal symptoms. The person trembles, sweats, vomits, and has cramps. Withdrawal from heroin is so painful that people will often do anything to get more of the drug.

Two more examples of narcotics are morphine and codeine. These drugs can be used legally if prescribed by a doctor. They are used to help relieve pain. But, many people abuse these drugs. When these drugs are taken without a prescription, they are considered illegal drugs.

Hallucinogens

Health Fact

A drug injected into the body takes about 15 seconds to get to the brain.

Hallucinogens are drugs that cause the abuser to experience sounds, sights, and feelings that are not real. LSD is one type of hallucinogen. It is often called *acid*. LSD is usually taken by mouth. A person on an LSD trip may "see" beautiful colors and creatures that are not real. He or she may also "see" a person turn into a horrible monster. Under LSD, a person cannot control his or her physical actions.

Another hallucinogen is PCP. It is often called *angel dust*. It can be smoked, injected, swallowed, or sniffed. Under the influence of PCP, a person moves slowly and hallucinates. Many PCP abusers become violent and uncontrollable. Because they do not feel pain on the drug, they may hurt themselves without knowing it.

Ecstasy is a widely abused drug. It is also known as *E*, and *XTC*. Ecstasy makes a person feel as if they have more energy. It causes a person to feel, see, and hear things differently. Ecstasy also can cause a person's body temperature to rise, and can cause damage to the liver, kidneys, and cardiovascular system.

Marijuana

Marijuana can act like a depressant or a stimulant. Marijuana comes from the dried, chopped tops or leaves of the hemp or marijuana plant. Its scientific name is *Cannabis sativa*. Marijuana is also called *pot* or *weed*.

Marijuana is most often smoked in a cigarette called a *joint*. The smoker usually gets pleasant, dreamy thoughts at first. Audio, sight, and time sensations all change. Some marijuana abusers do not communicate with others. Others laugh a lot and act silly.

Many marijuana abusers think this drug is less harmful than alcohol. After all, it does not leave you with a hangover. But, marijuana is a harmful drug. Studies show that marijuana causes a person to lose interest in school and work. Learning becomes difficult. The marijuana abuser has trouble remembering things, even when he or she is not taking the drug. Marijuana affects a person's ability to drive safely. Also, marijuana smoke contains chemicals that cause cancer.

 **Health Fact**

THC is the chemical in marijuana that causes a person to get "high." THC stays in the body for about 10 days.

Inhalants

Talk About It

A person can go into a coma or die after just one use of an inhalant. Over 1,000 teenagers die each year from using inhalants. How can you protect yourself and your friends from these risks?

Inhalants are substances that people abuse to feel high by breathing in the fumes of the substance. Some people abuse laughing gas, glue, spray paint, aerosol, and even gasoline as a drug. Most inhalants act in a way that is similar to alcohol. They slow down body functions. Abusing an inhalant can lead to nose bleeds, nausea, tiredness, and loss of coordination. Inhalants can cause serious medical problems such as liver and kidney damage, brain damage, and even death.

How Do Peer Influences Affect Drug Abuse?

Young people who abuse drugs or alcohol for the first time often do so because of pressure from friends or schoolmates. This is called *peer pressure*. People give in to peer pressure because they want to be accepted or because they do not want to be made fun of.

People can refuse peer pressure by saying no or by spending time with healthy, drug-free people.

Spending time with healthy, drug-free friends is a good way to avoid peer pressure about drug abuse.

Drugs Hurt Everybody

Some people say that taking drugs is a personal choice. They claim that it does not affect anybody but the drug abuser. However, drugs do affect others.

Drug abuse often leads to violence. Drug abusers are more likely to act impulsively. This impulsive behavior can cause the abuser to injure him or herself or others. Drug abuse has lasting negative effects on the abuser's family. There are also financial costs that go along with drug abuse. People may spend all of their money on drugs. Then, they may not be able to pay for food, clothing, or a home. Also, drug abuse counseling and treatment are usually very expensive.

Read the situations below. Think about how the drug abuser's behavior affects other people.

- A husband and wife argue over the husband's cocaine use. They eventually get divorced.

- A father spends all of his savings on drugs. His family cannot afford to pay rent and they are kicked out of their home.

- A woman abuses crack while she is pregnant. Her baby is born addicted to crack and is blind.

- A heroin addict holds up a grocery store in order to get drug money. During the robbery, two people are shot.

- Two gangs in the same neighborhood fight over the rights to sell drugs. The fight ends in violence.

- Some high school students abuse marijuana and then try to drive a car. They get into an accident and kill someone in the other car.

Write About It

One way that communities have tried to prevent drug abuse is by designating *drug-free school zones*. Anyone caught carrying, selling, buying, or using drugs in these areas, usually receives much greater penalties than in other areas. What are some other ways your community has tried to stop drug abuse? Write your answer on a separate sheet of paper.

Making Decisions and Getting Help

Have you ever been offered drugs? If you have not, you probably will be. Everyone must decide for themselves whether or not to abuse drugs.

When you make your decision, remember to look at your values and goals. What is important to you? Is feeling high for five minutes on crack more important than life itself? Are a few hours of pleasant thoughts on marijuana worth lessening your ability to learn and remember? How will your loved ones be affected by your drug abuse?

If you need more information or need help with a drug addiction, you can call the Drug Abuse Information and Referral Line 1-800-662-HELP. You can also talk to friends, teachers, parents, and counselors. There may be community health centers in your area that can help, too.

 Check Your Understanding

Write your answers in complete sentences.

1. What are the effects of heroin abuse?

2. How does marijuana affect learning?

3. List three ways drug abuse affects other people.

4. CRITICAL THINKING If you know someone with a drug problem who refuses to stop using the drug, what can you do to help him or her?

LIFE SKILL
Using Refusal Skills When Communicating

Most of the time, young people try cigarettes, alcohol, or drugs because other people tell them they should. These young people may have felt pressure to follow along.

You should not have to do something that you know can be harmful. But, saying no is not always as easy as it seems. Some people do not accept no for an answer. You can use refusal skills to help you say no.

You can always say no to peer pressure.

Practicing saying no can be helpful. Say it over and over in your mind. Practice with a friend or family member.

REFUSAL SKILLS

When you are feeling pressured, try the following steps:

- Say no and mean it.
- Give a reason for saying no.
- Offer an alternative action.
- Walk away from the situation if necessary.

Applying the Skill

Think about this situation. You are at a party. Some people are smoking marijuana. A friend offers you a smoke. Work with a partner to write a dialog for this situation.

What are some reasons you could give for saying no to tobacco, alcohol, and other drugs?

Summary

All drugs can be abused. Some reasons people abuse drugs are peer pressure, curiosity, the need to treat pain or illness, to forget about troubles, and to have fun.
The four main types of drugs are stimulants, depressants, narcotics, and hallucinogens.
Drug abuse can lead to serious health problems. It can also cause psychological dependence.
Cocaine is a dangerous, addictive drug. Cocaine abusers risk many health problems, including heart attacks. Crack is a cheap, powerful form of cocaine.
LSD, PCP, and ecstasy are hallucinogens. A person on these drugs may hear and see things that are not real.
Marijuana can cause a person to lose the ability to learn and remember. It can also cause cancer.
Drugs affect everybody. It is important to think about your values, your future goals, and your loved ones before you decide to abuse drugs.

addictive

barbiturate

hallucinogen

psychological
dependence

tranquilizer

Vocabulary Review

Complete each sentence with a term from the list.

1. A drug that slows down the nervous system is called a _____.

2. A drug that has a calming effect is called a _____.

3. a person's emotional need for a drug is called _____.

4. A drug that causes a person to see, hear, and feel things that are not real is called a _____.

5. Something that is habit forming is called _____.

Chapter Quiz

Write your answers on a separate sheet of paper. Use complete sentences.

1. Give three reasons why people abuse drugs.

2. Give an example of how a drug can be both used and abused.

3. What are the four main types of drugs? Give an example of each.

4. What do amphetamines do to the body?

5. Why do cocaine and other stimulants cause heart attacks?

6. Why is PCP so dangerous?

7. What are the effects of marijuana?

8. How do inhalants affect the body?

9. How do drugs affect the community? Give at least two examples.

10. Name one way you could get more information about drug abuse.

CRITICAL THINKING

11. Describe the ways that drug abuse affects your physical, emotional, and social health.

Online Health Project

Use the Internet to research drug addiction. Find out what causes the addiction, how the addiction affects the drug abuser and others, and how people fight the addiction. You may also want to describe the experiences of someone who once was addicted to the drug as an example.

HEALTH LINKS.
Go to www.scilinks.org/health.
Enter the code **PMH350** to research **drug addiction**.

Unit 4 **Review**

Comprehension Check

On a separate sheet of paper, write how each of the things below will benefit your health.

1. giving up cigarettes
2. never smoking crack
3. following the directions on medicine packaging
4. avoiding alcohol
5. finishing all of your antibiotics

Analyzing Cause and Effect

Write a sentence or two explaining what might cause the following. The information you learned in Unit 4 will help you.

6. A pot smoker begins to lose his memory.
7. A PCP user jumps through a glass window and feels no pain.
8. A cigarette smoker gets bronchitis several times a year.
9. A person gets sick after taking someone else's prescription medicine.
10. A teenager falls down after one use of an inhalant.

Writing an Essay

Answer the questions below on a separate sheet of paper. Use complete sentences.

11. Suppose you want your friend or family member to stop smoking. What negative health risks can you talk about to try to convince him or her to quit?
12. What is the difference between over the counter medicines and prescription medicines? Are either of these drugs completely safe?
13. What are some of the short-term and long-term effects of alcohol use?
14. How does alcoholism affect people other than the alcoholic?
15. List the four main types of illegal drugs and give an example of each.

Managing Your Health

Teenagers are often pressured into using drugs or alcohol. Consider your own personality, peer relationships, and values. Describe a specific situation in which drugs or alcohol are offered to you. List some refusal skills you can use to handle this peer pressure in a healthy way.

Family Living and Sexuality

Before You Read

In this unit, you will learn about being part of a family and becoming an adult. You will learn how your body develops and matures. You will also learn how sexuality plays a role in your health and well-being.

Before you read, ask yourself the following questions:

1. What do I already know about family living and sexuality?

2. What questions do I have about how my body develops and changes as I get older? How do these changes affect my health?

3. How should I react to my sexuality and that of others? How do the decisions I make affect my health?

Families support each other and give a sense of belonging. This support helps your social and emotional health.

Learning Objectives

- Describe different types of families.

- Explain how family relationships affect your health.

- Describe the roles of parents and children in a family.

- Give some examples of conflicts in a family.

- LIFE SKILL: Analyze the relationship between family and health.

Chapter 17 ▶ Family Living

Words to Know

family	a group of people that is related by genetics, marriage, or legal action
nuclear family	a family made up of two parents and one or more children
extended family	a family made up of a nuclear family and other relatives, such as grandparents, aunts, uncles, and cousins
separation	an agreement between a married couple to stay married but live apart
divorce	a legal end to a marriage
custody	a legal arrangement of who is responsible for the care and well-being of a child
guardian	a person who cares for a child but who is not the parent

Different Kinds of Families

A **family** is a group of people that is related by genetics, marriage, or legal action. A **nuclear family** has two parents and one or more children. Sometimes a family has only one parent, either a mother or a father, and one or more children. This kind of family is called a *single-parent family*.

Today, families come in all different sizes and variations. Some married adults do not have any children. But, they are still a family—a family of two. Some families have same-sex couples.

Grandparents, aunts, uncles, and cousins are part of your **extended family**. Sometimes all the members of an extended family live together. Usually, grandparents, aunts, and uncles live in their own homes.

Family Relationships and Your Health

Talk About It

What kinds of jobs do you have to do around your home? How do you think doing these jobs helps your family?

Family members support each other in many ways. No matter what type of family you have, the relationships with your family members are very special. Your family gives you a feeling of belonging and acceptance. Your family can make you feel better about yourself.

If you are close with your family, you always have someone to talk to. You can share your hopes and dreams. You can also share your troubles and problems. Maybe you have had a bad day at school or a fight with your best friend. Talking with your family can help you feel better. When you are close with members of your family, you feel good about yourself. You also have better emotional and social health.

Roles Within the Family

Each member of a family has certain roles, or jobs, to do. Your parent or parents have many important family jobs. Parents make sure the family has a safe place to live. They make sure the family has food to eat and clothes to wear. Parents take care of the family when someone is sick or in an emergency.

Children in a family also have roles. Many children have jobs to do around the home. In many families, children must keep their rooms clean. In some families, teenagers have the job of doing their laundry. Younger children may have to put away their clean clothes.

Doing laundry is one role that children have in some families.

Conflicts in a Family

Not all families get along at all times. Usually, family arguments or fights are resolved and the family continues to function together. But, sometimes families cannot solve their problems. These families can become *dysfunctional*.

Parents and teenagers can solve almost all disagreements through cooperation. The best kind of cooperation is talking with each other. Good communication skills will keep your family relationships healthy.

Parents can also have conflicts with each other that have nothing to do with the children. Sometimes, married couples live apart because they are having problems in their marriage. A **separation** happens when married people agree to stay married but live apart. During a separation, a couple usually tries to solve their problems. They often work with a person called a counselor to save their marriage.

Write About It
What are some of the things that you and your parents do not agree on? How does this affect your relationship?

Remember
A counselor is someone who helps people with their problems.

People in Health

MICHAEL BARBERO—MARRIAGE COUNSELOR
Problems or issues sometimes occur in relationships. Couples may decide to get outside help from a marriage counselor.

Michael Barbero is a family and marriage counselor. He has a master's degree in social work, and works with clients between 30 and 35 hours each week.

Counseling can help couples solve problems or issues in their relationships.

Michael uses communication skills during sessions with his clients. He speaks with clients in both individual and joint sessions. Cultural, ethnic, and economic backgrounds are discussed. Finding out as much about a couple as possible helps a marriage counselor better understand the relationship. Michael feels that the best part of his job is helping others. After working successfully with a couple, it is satisfying to see a couple begin to feel positive about their relationship.

CRITICAL THINKING Why would it be important to discuss cultural and ethnic backgrounds with a couple?

Divorce

Health & Safety Tip

If your parents are getting a divorce, remember two things: First, you are not alone. Many teenagers have felt or are feeling the same things as you. Second, remember that children are not responsible for their parents' divorce.

Some married couples cannot work out their differences. They decide to end their marriage. A **divorce** is the legal end to a marriage. If the couple has no children, they divide their property—their home, cars, and other similar things. Then, they live their own lives.

If children are present, one or both parents will have custody of the children. **Custody** is the legal arrangement of who is responsible for taking care of a family's children and making decisions for them. Sometimes, the children live with only one parent. Other times, the children live with both parents at different times of the year. Sometimes, the children live with a guardian. A **guardian** is a person who cares for a child but is not the parent. A guardian can be a relative, a friend, or a foster parent.

Separation and divorce are not happy events. They create a great amount of stress within a family. Each family member is affected in some way. The best way to understand what is happening is to talk with your parents or guardian. Ask them questions about what will happen to you and to them. Tell them how you feel. Let your parents know if you are unhappy or angry. Tell your parents if you feel guilty. Talking to one or both of your parents will help you through these difficult times. It is healthy for people to express their feelings.

✓ Check Your Understanding

Write your answer in complete sentences.

1. What are three types of families?

2. What are three things parents and teenagers may not agree on?

3. CRITICAL THINKING How can having dinner all together improve the health of the family?

LIFE SKILL
Relating Family Health and Personal Health

The health of your family relationships can affect your health overall. Everyone has some stress in his or her life. Stress is normal. If you have good relationships with your family members, they can help you deal with that stress. Poor family relationships can make stress much worse and make you sick.

Very often, children blame themselves when their parents divorce, even though it is not their fault. Too much stress at home can also lead to depression.

Problems in your family relationships can affect your social well-being. If a problem at home has you feeling nervous or upset you might act mean to a friend. If you do not want to talk about problems at home, you might avoid your friends.

Good friends will help you during difficult times. Talking to friends, teachers, counselors, a school nurse, or a doctor can help you cope with the stress at home. Your health may depend on it.

An unhealthy family relationship can affect your school work or your job.

Answer the questions below.
1. What types of health can be affected by poor family relationships?

2. What are some things you could do if a problem relationship at home were affecting your relationships with your friends?

CRITICAL THINKING Write a paragraph to describe how family relationships—either positive or negative—can affect your daily life.

Applying the Skill
Sam's parents argue a lot. They say their arguing has nothing to do with him. On a separate sheet of paper, explain how Sam's parents are wrong. Explain how their fighting can affect Sam's daily life and his health.

Summary

There are many types of families today. Some types of families include nuclear families, single-parent families, and extended families.

Having healthy relationships with your family has positive effects on your own social and emotional health. Communication is the best way to maintain healthy family relationships.

All families go through a conflict at some time. Parents and children may not agree. Parents may also have personal conflicts with each other. These conflicts may lead to separation or divorce. It is important for children of divorced parents to understand that the divorce is not the children's fault.

custody

divorce

extended family

family

guardian

nuclear family

separation

Vocabulary Review
Complete each sentence with a term from the list.

1. A family made up of a nuclear family and other relatives, such as grandparents, aunts, uncles, and cousins, is called an _____.

2. An agreement between a married couple to stay married but live apart is called a _____.

3. A family made up of two parents and one or more children is called a _____.

4. The legal end to a marriage is called a _____.

5. The legal arrangement of who is responsible for the care and well-being of a child is called _____.

6. A _____ is a group of people that is related by genetics, marriage, or other legal action.

7. A person who cares for a child but who is not the parent is a _____.

Chapter Quiz

Write your answers on a separate sheet of paper. Use complete sentences.

1. What is a family?

2. What are three types of families?

3. What family members are part of an extended family?

4. If you live with your father only, what type of family do you belong to?

5. In what ways does your family support you?

6. How does having strong family relationships affect your health?

7. How can you keep your family relationships healthy?

8. What is your role within your family?

9. What are two types of conflicts within a family?

10. How is a separation different from a divorce?

CRITICAL THINKING

11. List three things that you think are necessary for a family to be healthy.

12. Suppose a friend of yours has parents who are getting divorced. What can you do to help this person?

Online Health Project

Families today are different than they were 50 or even 20 years ago. Find out more information about these "new" types of families. Explain the roles of the individuals in each family. Describe how these families are similar to and different from your family. Write a report that summarizes your findings.

HEALTH LINKS.

Go to www.scilinks.org/health. Enter the code **PMH360** to research **family**.

The teenage years are a time of changes, accomplishments, and new responsibilities.

Learning Objectives

- Explain why people develop differently.
- Describe the physical changes that people go through during their teenage years.
- Describe ways in which people change emotionally and socially during their teenage years.
- Explain why making wise decisions is important.
- **LIFE SKILL:** Set a goal for getting and keeping a part-time job.

Chapter 18 ▶ Development and Maturity

Words to Know

adolescence	the teenage or young-adult years
growth spurt	a time when a person grows more quickly than at other times
pituitary gland	a small gland in the brain that controls the release of hormones
peer pressure	pressure from friends to act or think in a certain way

Changes in the Young-Adult Years

The teenage, or young-adult, years are called **adolescence**. This time is often confusing for people. Your body is changing. Your emotions are changing. You are faced with becoming an adult. Of course, you cannot wait for adult responsibilities. Working, driving, and dating can all be exciting. They can also be stressful. What is happening to the body and mind during these young-adult years?

Talk About It

In what ways does your life change as you become a young-adult?

Physical Development

The biggest change during the young-adult years comes with puberty. You read in Chapter 5 about the ways the body becomes sexually mature. Puberty can be an awkward time for some people. The reproductive organs are not the only parts of the body that are affected. The skin may become more oily, which produces acne, or pimples. Hair grows on the body. In males, puberty causes the voice to change. In females, the breasts develop and the hips may become broader. These physical changes can make a young person feel uncomfortable. But, these changes are a natural part of growing up.

Remember
Puberty is the time when physical changes occur in the young-adult body.

Other changes happen during the young-adult years, too. Most noticeably, they have to do with your height and weight. Each person grows at a different time and a different speed.

There is a great variety of heights among teenagers of the same age.

Different Growth Rates

Most girls begin growing and developing about two years before most boys. Even among girls, there are great differences in how and when they grow.

Sometimes, a person grows more quickly than at other times. This type of growth is called a **growth spurt**. Growth spurts can happen several times during the teenage years.

A Wide Range of Heights and Weights

There is a wide range of heights and weights for each age group. No chart or table can tell you exactly the right height and weight for your age. The tables on page 255 shows some typical heights and weights for people who are 12 to 15 years of age.

Remember
Heredity is the passing of traits to you by your parents.

Different growth rates, growth times, and body builds are caused by heredity. The traits that control your growth and adult size come from your parents.

GIRLS		
Age	Height	Weight
12	53–65 in.	60–140 lbs.
13	55–67 in.	66–155 lbs.
14	58–68 in.	74–165 lbs.
15	59–70 in.	83–172 lbs.

BOYS		
Age	Height	Weight
12	53–68 in.	65–140 lbs.
13	54–70 in.	67–161 lbs.
14	56–73 in.	73–178 lbs.
15	58–74 in.	81–192 lbs.

The Pituitary Gland

The brain also helps control growth. The release of your body's hormones is controlled by the **pituitary gland**, which is found in the brain. This gland sends hormones into your body. The hormones cause growth spurts, especially during the puberty years.

Write About It

What is the normal weight range for a 15-year-old girl? Are you surprised at the range?

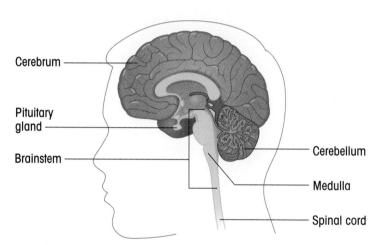

Cerebrum

Pituitary gland

Brainstem

Cerebellum

Medulla

Spinal cord

The brain is the control center of the body.

✓ Check Your Understanding

Write your answers in complete sentences.

1. What are three physical changes that happen during adolescence?

2. What is a growth spurt?

3. CRITICAL THINKING What is the role of the pituitary gland in growth?

Emotional and Social Development

Think About It

Think of three clubs or groups at school you might enjoy being part of.

Development in the young-adult years is physical and emotional. It is a time of changing feelings and new responsibilities. You will notice these changes most often in your relationships with the opposite sex, your friends, and your parents.

The Opposite Sex

Interest in the opposite sex begins at different times for different people. Most girls begin to mature physically about a year or two before boys do. That is why in a seventh or eighth grade class, most boys are not as interested in girls as girls are in boys. In a few years, the boys catch up.

Joining mixed groups is a good way to get comfortable with the opposite sex.

For most people, feeling relaxed around the opposite sex takes practice. It helps to become friends with people of the opposite sex before dating. You can practice talking and being yourself. Also, you can learn how to care about others without the pressure of dating.

Dating can help you become closer to someone. You can find out more about that person's likes and dislikes. Dating can also be a way to find out more about yourself. Different people are ready to begin dating at different times. If you do not feel ready to date, there is no reason to start. If you do decide to date, it is important to keep a few things in mind:

- Go out as a group date first to get used to dating.
- Communication is very important. You should be able to talk about your feelings with your partner.
- Sexual contact too soon can result in an unplanned pregnancy or a sexually transmitted disease.
- Let someone know where you will be, carry a mobile phone with you, and avoid alcohol.
- Breaking up can be very difficult for young-adults to cope with. Talking about your feelings may help.

Think About It

WHAT ARE SOME IMPORTANT QUESTIONS ABOUT DATING?

A date should be an enjoyable time. But, some dates turn into bad experiences. One person may try to control the other or become physically violent or abusive. Research shows that this problem occurs in one out of every three high school relationships.

Respect and responsibility are two parts of a healthy relationship.

You can protect yourself from dating violence. Spend time with a person as part of a group. Notice whether he or she shows respect for you and others. Think about whether the person shares your values. If not, avoid dating him or her. Saying no to a date with the wrong person shows that you value your health and safety.

YOU DECIDE While on a date, your partner suddenly begins to make fun of you and put you down. When you ask your partner to stop acting this way, he or she becomes angry. What should you do?

Write About It

What if you disagree with something a friend is doing? How can you speak up without hurting the friendship? Write your answer on a separate sheet of paper.

Friends

Perhaps you are not as close to some of your old friends now as you used to be. This is a natural part of growing up—as you lose some friends, you gain new ones.

Having friends is important. Sharing questions and talking about ideas with friends can be a valuable experience. Out of this sharing can come answers to many questions you may have about life and its problems.

Having a close group of friends can give you a protected, secure feeling. You may feel great loyalty to your friends. It is the same kind of loyalty that holds families and other adult groups together.

But, any group can sometimes get in the way of your own ideas, feelings, and thoughts. **Peer pressure** can make people act or think in a certain way just to belong.

Often, a person who gives in to peer pressure is afraid of losing his or her friends. Peer pressure can get a person involved in smoking, drinking, drugs, crime, and prejudice. Remember that no one can force you to act in a way that you believe is wrong. At some time, you may find yourself being pressured by your peers. At other times, you may be the one pressuring others to do as you do.

Coping With Your Changing Relationships

How can you cope with changes in relationships? How do you stand up to your peers? How can you let your parents know that you are ready for responsibility? There are two important keys:

- Learn how to communicate well.
- Learn how to make good decisions.

Talk About It

What does the word *respect* mean to you? How do you want people to show respect? How do you show respect to others?

Communicating effectively is not always easy. It is easy to become angry or to be too shy to speak up. Many times, anger or fear get in the way of saying how you really feel. Chapter 20 will give you a look at how to communicate successfully.

By making good decisions before you act, you will find it easier to not be pressured by peers. You may also get along better with your parents. Little by little, as parents see that you think and act responsibly, they will begin to relax. They will feel more comfortable letting you be independent.

Parents

Making the change from childhood to adulthood is not easy—for you or your parents. Many of the changes you are making may be a shock to your parents. That is because they still think of you as a child. As a child, you depended on them for everything. They were used to making decisions for you. Now, suddenly, you want to make your own decisions. This change often ends in anger and misunderstanding. It is important to talk to your parents to gain their respect.

Talk About It

What rules do you have at home? Do you think they are fair? Why or why not?

New responsibilities, such as driving, can be an opportunity to gain your parent's respect.

How to Make Wise Decisions

Write About It

Make a list of some of your values. Think about how you use these values in decision making.

1. Start by looking at your values. Values are the attitudes and beliefs that are most important to you. Honesty, keeping healthy, spending money wisely, and being nice to people are all examples of values.

2. When you have to make a decision about something, take the time to think it over carefully. Never let anyone make you feel you have to decide something right away. If someone says, "Decide now or never," "never" is usually the better choice.

3. Gather information about the possible consequences of your decisions. Suppose your friends want you to skip a class. Think about how this will affect your grades and future. Think about how your family and teachers will feel.

4. Decide which choice is better. Whatever decision you make, you will be giving something up. If you skip the class, you may be giving up your future and your family's trust. If you do not skip the class, you may lose a few friends. You must be prepared in both cases.

5. Before you make your choice, go over your values one more time. They will help you make the decision that is right for you.

✓ Check Your Understanding

Write your answers in complete sentences.

1. How do your relationships with the opposite sex change during adolescence?

2. What is peer pressure?

3. How can you get your parents to give you more freedom and independence?

LIFE SKILL
Setting a Job Goal

Many teenagers have a part-time job. A part-time job is one that you do a few hours each week. You can earn money with a part-time job. You also can learn new skills.

You can take action to succeed at a part-time job. For example, you can set a goal to succeed at a job of walking dogs. To reach that goal, you need to make an action plan. The list below shows the steps for making an action plan.

Many teens earn extra money with a part-time job, such as walking dogs.

Action Plan for a Job Goal

STEP 1 **Write down your goal.** "I want to succeed at my job of walking dogs."

STEP 2 **List steps to reach the goal.** One step may be telling people that you can walk their dogs and making a schedule.

STEP 3 **Set up a timeline.** Decide how much time each step should take. Keep track of each step on your calendar.

STEP 4 **Identify any obstacles.** List problems that might make it difficult to reach your goal. Think of ways you might overcome them.

STEP 5 **Identify sources of help.** Decide who or what could help you reach your goal.

STEP 6 **Check your progress.** Review your plan. If your plan is not working, make changes so you can still reach your goal.

After two weeks, you find that your plan has a problem. You do not have enough time for friends and homework. What change could you make?

Applying the Skill

Choose a part-time job that interests you, or your current job if you have one. Write an action plan for the job on a separate sheet of paper. Use the steps to help you.

Summary

The teenage, or young-adult, years are times of both physical and emotional change. Puberty is a time that includes many physical changes.

Height and weight vary greatly among teenagers. Teenagers grow and develop at different rates because of heredity.

Changing emotions and needs also change relationships. Teenagers may find themselves thinking and feeling differently about the opposite sex, friends, and parents.

Making good decisions is an important way to cope with new feelings, conflicts, and responsibility. To make good decisions, a person must be in touch with his or her values.

adolescence

growth spurt

peer pressure

pituitary gland

Vocabulary Review

Complete each sentence below with a term from the list.

1. A small gland in the brain that controls the release of hormones is the _____.

2. A time when a person grows more quickly than at other times is called a _____.

3. Pressure by friends to act or think in a certain way is known as _____.

4. The teenage or young-adult years are also called _____.

Chapter Quiz

Write your answers on a separate sheet of paper. Use complete sentences.

1. Name three physical changes that happen to teenagers.

2. What does the pituitary gland have to do with your development?

3. A girl is 14-years-old and weighs 130 pounds. Is this a normal weight? Explain your answer.

4. What are three relationships that may change during adolescence?

5. What is the benefit of going out with group of mixed friends before dating?

6. Give an example of peer pressure.

7. What are three common causes of conflict between parents and teenagers?

8. What are values?

CRITICAL THINKING

9. Why is it important to learn appropriate ways to express anger or fear?

10. How can making good decisions help earn you more responsibility and independence?

Online Health Project

Teenagers are often very concerned about their physical appearance. They spend a great amount of time worrying about their body, hair, and clothing.
How do you think your body image affects your emotional and social health? Why do you think teenagers are so concerned about body image? Write an essay describing your ideas. Use the Internet to find facts about body image and teenagers.

HEALTH LINKS℠
Go to www.scilinks.org/health. Enter the code **PMH370** to research **self-image**.

Making healthy choices about sex can lead to a long, satisfying relationship.

Learning Objectives

- Define sexuality and name some sexual issues.

- Explain the responsibilities and consequences that come with being sexually active.

- Describe the problems and responsibilities that come with teenage pregnancy.

- Explain why abstinence is a smart choice.

- LIFE SKILL: Make healthy choices about sex.

Chapter 19 ▷ Sexuality

Words to Know

sexuality	the state of being sexual; a person's sexual interests and issues
heterosexual	a person who is attracted to those of the opposite sex
homosexual	a person who is attracted to those of the same sex
abstinence	deciding not to have sex and sticking with your decision

What Is Sexuality?

Sexuality means the state of being sexual. It covers a whole range of sexual interests and issues. It has to do with who you are and what you think about sex.

Your sexuality starts when you are born a male or a female. Since you were little, you probably noticed some basic differences between boys and girls. You learned about these differences in Chapter 5. There are also differences in the emotions and personalities of men and women. Scientists argue about these differences. Some say that differences between boys and girls are inborn traits. Others say these differences are learned. They say that if they were treated the same way, boys and girls would act alike.

During late childhood and early teens, sexuality becomes a bigger part of a person's life. This process begins with puberty. Sexuality continues to change throughout life.

Puberty is often the time at which boys and girls become attracted to one another.

Write About It

Has your family given you rules about dating? Do you think the rules are fair? Why or why not?

When Attraction Begins

During puberty, our bodies begin to mature and develop. Sexual feelings may start to happen. In most cases, a young man will begin to have sexual feelings about women. Young women are usually attracted to men. People who are attracted to the opposite sex are called **heterosexuals**.

A smaller percentage of the population is attracted to people of the same sex. These people are called **homosexuals**. Men who feel sexually attracted to men are often called *gay*. Women who feel sexually attracted to women are called gay or *lesbian*.

No one is really sure what causes a person to be homosexual. Some think sexual preference has to do with early childhood relationships. Some think that hormones are responsible. Others think homosexuality is inherited, like eye or hair color. No one knows for sure. It is believed that about ten percent of the American population is homosexual.

Dating

During the early teen years, you may begin to date. Some young people are very comfortable with dating. Others find dating awkward and stressful. Much of the stress a person feels is brought on by these kinds of worries:

- Will the other person like me?
- What will we talk about?
- How do I keep from sounding stupid?

Remember, the person you are dating probably has these same kinds of worries. The worries become fewer as you spend more time dating.

Dating Exclusively

Sometimes young men and women decide to date exclusively. This means that they date only each other. You can learn a lot about relationships when you only date one person. Together, you can learn how to cooperate and compromise. You can learn how to get along with each other's friends. You can learn how to deal with different interests and needs at the same time.

✓ **Check Your Understanding**

Write your answers in complete sentences.

1. How are teenage men and women different?

2. Do you think dating exclusively is a good idea? Why or why not?

3. If you were a parent, what rules would you give your daughter or son about dating?

Think About It

HOW ARE SEXUALITY AND ADVERTISING RELATED?
It is estimated that an average female has observed about 250,000 commercials by the time she is 17 years old. Many of those ads use sexuality to sell products. The ads try to convince people that using a product will increase their attractiveness.

Some experts believe the ads send unhealthy messages. They say the ads make it seem that how you look determines self-worth. The experts believe the ads encourage viewers to practice unhealthy behaviors such as diets that may not work and using steroids.

YOU DECIDE Think about the last three commercials you observed. How did the ads use physical appearance to convince viewers to buy the products?

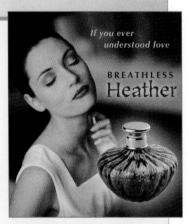

This ad suggests this perfume will make you attractive.

Making Decisions About Sex

Talk About It

Many men and women do not feel comfortable talking about sex. Is this attitude healthy?

Someday you will have to decide whether you should have sex or not. Perhaps the person you are dating will pressure you. Maybe you will be curious, or your friends will think you are not "grown up" if you do not have sex.

Your body is ready to have sex long before your emotions are ready for it. When two young people have sex, one or both may feel bad about it, even if they are in love. By waiting until you are older, you will know yourself and your feelings better. You will have a better ability to cope with sexual and emotional feelings.

What Is Safe Sex?

Health Fact

Researchers are finding that AIDS is spreading faster among teenagers and young-adults than any other group. In 2002, there were 3.2 million teens under the age of 15 living with AIDS.

You have read about HIV infection and other sexually transmitted diseases (STDs). Some of these diseases, such as HIV (which leads to AIDS), can kill you. For that reason, practicing safe sex can save your life. The following guidelines tell more about safe sex.

- The safest sex is no sex at all. Some people choose to put off being sexually active until they are more mature and have more information about themselves and their potential partners. Choosing not to have sex is the healthiest choice.

- If you choose to have sex, the male partner must wear a condom. This will help protect both people from HIV and other STDs.

- One way that the HIV infection and AIDS is spread is through sex, either heterosexual or homosexual. In fact, most people with HIV got it through heterosexual sex.

- The more sexual partners you have, the higher your risk of spreading or getting HIV or other STDs.

- Communicate with your partner. If you think you have an STD, get medical help. Also, tell anyone you have had sex with about the disease. This may be a very hard thing to do. But you may be saving a life by doing it.

- HIV and AIDS can be given to an unborn child by the child's mother. Unsafe sex not only can be a risk to you—but it can be a risk to a baby if you have one.

Remember

AIDS is a disease that destroys a person's immune system. There is no cure for AIDS.

Teenage Pregnancy

Sex gives people pleasure. It can be a way of showing love. But, sex can also lead to pregnancy. Deciding to have sex means taking on responsibilities, for both men and women.

Are you ready to have a baby? Are you willing to get up in the middle of the night to feed it? How will you feel when you cannot go out with your friends because there is no one else to care for the baby? How will it feel when your dreams for a career must be put off for years because you have a baby? Having a baby takes a lot of responsibility and sacrifice. Having a baby is also very expensive.

Having a baby is a big responsibility.

Practicing Abstinence

Your values and goals should guide you when you are deciding about sex. Talk to an adult when you have questions. With enough valid information, you should be able to make a decision that is right for you.

Many teenagers today are deciding to not have sex until they are married. Not having sex and sticking with that decision is called **abstinence**. Abstinence will prevent unplanned pregnancies, sexually transmitted diseases, and emotional problems. Practicing abstinence is not an easy goal. Below is a list of tips you can use for sticking with your goal:

- Consider your values, priorities, and future goals. Think about where you want to be in five years. Then, think about how sexual activity may affect these goals.

- Talk about your decision with your partner. If you cannot talk about it with your partner, then you are probably not in a healthy relationship.

- Set limits for how you and your partner can show your feelings. Avoid being in situations in which you may not be able to stay within these limits.

✓ **Check Your Understanding**

Write your answers in complete sentences.

1. What are two consequences of sexual activity?

2. What is the only way to avoid sexually transmitted diseases completely?

3. CRITICAL THINKING Why is good communication important in a dating relationship?

LIFE SKILL
Making Healthy Choices About Sex

Making healthy choices about sex means doing what is best for yourself. You need to think about how you and your partner will be affected. Having sex may expose you to a sexually transmitted disease. There are also emotional risks. When two young people have sex, one or both may feel bad about it. Talk to your partner about your feelings, concerns, and values.

Making a Decision About Sex

STEP 1 Identify the decision you need to make.

STEP 2 List your choices.

STEP 3 Cross out choices that are harmful or might go against your beliefs.

STEP 4 Think about the possible results of the remaining choices.

STEP 5 Select the best choice.

STEP 6 Explain how you would carry out that choice.

STEP 7 Describe the possible results of your choice.

John decides to tell Laura that he does not want to have sex. But, he does not want to end their relationship. Write a paragraph that tells John the best way to tell Laura "no sex" without hurting her feelings.

The decision-making model can help you make good decisions about sex.

Applying the Skill
Think about the situation where John decides not to have sex with Laura. Use the decision-making model to show how he might have come to his decision.

Summary

Sexuality has to do with sexual interests and issues. Sexuality begins at birth. It is shaped through your life by learning and by society.

Dating is a way to become more comfortable with yourself and others. By dating, you can meet and learn about many different types of people.

With sex comes responsibilities. Teenage pregnancy is one risk a couple takes when they have sex. Catching and spreading STDs are other risks. Upset and hurt feelings can also result from having sex.

An HIV infection can be prevented by practicing "safe" sex. Abstinence is the best way to avoid all problems with sex.

abstinence

homosexual

heterosexual

sexuality

Vocabulary Review

Complete each sentence with a term from the list.

1. A person who is attracted to those of the opposite sex is _____.

2. The state of being sexual; a person's sexual interests and issues is called _____.

3. A person who is attracted to those of the same sex is _____.

4. Deciding not to have sex and sticking with your decision is called _____.

Chapter Quiz

Write your answers on a separate sheet of paper. Use complete sentences.

1. Is everybody's sexuality the same?

2. What makes men and women different?

3. What are two good reasons for dating exclusively?

4. What are two good reasons against dating exclusively?

5. Why is it better to wait until you are emotionally mature to have sex?

6. What are two ways that your lifestyle will be changed by having a baby?

7. What are two negative outcomes of having sex?

8. What can be done to prevent the spread of STDs?

9. What is the only sure way to avoid getting STDs?

10. What are the responsibilities of an STD-infected person?

CRITICAL THINKING

11. Suppose your boyfriend or girlfriend wants to have sex but you do not. What should you do?

12. Why do you think screening tests for STDs and AIDS are important to public health?

HEALTH
LINKS™

Go to www.scilinks.org/health. Enter the code **PMH380** to research **sexually transmitted diseases**.

Online Health Project

Sexually transmitted diseases are a negative consequence of having sex. Find out more information about STDs and teens. What are the trends for teenagers and these diseases? How can the diseases be prevented? Write a report that summarizes your findings.

Unit 5 Review

Comprehension Check

On a separate sheet of paper, give an example of each of the following types of families.

1. a nuclear family
2. a single-parent family
3. a same-sex family of two
4. an extended family
5. a family of two

Analyzing Cause and Effect

Write a sentence or two explaining why the following things might happen. The information you learned in Chapters 17–19 will help you.

6. A family becomes dysfunctional.
7. A mother has custody of her daughter.
8. A boy grows more quickly than he ever has before.
9. A boy develops an interest in girls.
10. A girl buys some perfume after seeing an ad in a magazine.

Writing an Essay

Answer the questions below on a separate sheet of paper. Use complete sentences.

11. What is a separation? How could a separation happen?
12. How is the pituitary gland related to puberty?
13. How can you cope with changes in a relationship?
14. What is sexuality?
15. What is safe sex? Why should people practice safe sex?

Managing Your Health

Name a decision that you or someone else might have to make. List the steps to making wise decisions. Next to each step, write how you would apply the step to your decision. Then, write about how using these steps helped you to make a wise decision.

Health and Society

Chapter 20
Communication and Relationships

Chapter 21
A Healthy Community

Before You Read

In this unit, you will learn how communication and friendship are related to health. You will also learn how to handle peer pressure in a healthy way. Finally, you will learn about healthy and unhealthy communities.

Before you read, ask yourself the following questions:

1. What do I already know about how good communication and friendship are related to health?

2. What questions do I have about what makes a community healthy?

3. How should I handle peer pressure?

Healthy communication means talking about little problems before they become big ones. At home or with friends, it is important to say what you mean.

Learning Objectives

- Describe the connection between communication and health.

- Explain how to communicate effectively.

- Describe some communication problems.

- Describe the signs of a healthy relationship.

- Explain the role of friendship in your health.

- Describe how you can handle peer pressure in a healthy way.

- **LIFE SKILL:** List several ways you can resolve conflicts.

Chapter 20 ▶ Communication and Relationships

Words to Know

passive	to be inactive or unable to say what you think, want, or need
aggressive	to act in a pushy, dominating way
assertive	to act in a positive, healthy way
body language	the ways you use your body to communicate how you feel or what you think
conflict	a disagreement between two or more people
gossip	comments made about another person without supporting evidence

Nothing Is Wrong

Kris and James are sitting on the school steps. They have been dating for three months. When James tries to take Kris's hand, she pulls it away.

"What is wrong?" asks James.

"Nothing," says Kris, looking down.

"Come on, something is wrong."

"I said nothing is wrong," Kris says again, quietly.

James rolls his eyes in disbelief. He feels helpless and angry.

"Call me when you are ready to talk," he says, and walks away.

Kris begins to cry when she sees him leave.

Are Kris and James communicating very well? What could they have done differently?

Talk About It

How do the body movements of Kris and James help communicate what they are not saying?

Communication and Health

Write About It

When was the last time you were misunderstood? Write a few sentences about the situation.

Communicating is giving and receiving information. What happens when you communicate poorly? Like Kris and James, have you ever been unable to communicate? You get stressed. Sometimes, people get so stressed that they get into fights. Other times they get frustrated and withdraw. Communicating poorly can lead to poor physical and mental health.

Communicating well is an important part of being healthy. Here are some things that good communicators can do:

- They say what they are thinking and feeling and get people to listen.
- They do better at their jobs.
- They have better relationships with people.
- They feel good about themselves.

Talk About It

Almost half of all marriages in the United States end in divorce. How might this be related to poor communication?

Part of the way you communicate is hereditary. Your temperament and personality may affect how calm or quick-tempered you are. But, communication skills are also learned. You learn your communication style from your family and friends through everyday experiences. But, whatever you inherit or learn can always be improved.

You start learning to communicate when you are very young.

Three Communication Styles

There are many different styles of communication. Three main styles are passive, aggressive, and assertive.

Passive people are usually inactive or unable to say what they think or want. Passive people are often withdrawn and shy. They often look down, talk softly, and slump over. Passive people have a hard time saying what is on their minds.

Aggressive people act in a pushy, dominating way. Aggressive people are the opposite of passive people. They often speak loudly. They may point their fingers, or fight when they get angry. Aggressive people may get what they want. But, they do so in an unhealthy way. Aggressive people have a hard time getting along with others.

Assertive people act in a healthy, positive way. They stand and sit tall. They speak in strong, clear voices. They say what is on their minds. But, they are willing to listen to the other people's thoughts and opinions.

People use different communication styles at different times. You may be assertive at home with your family. While on a date, you may be more passive. Your style of communication often depends on whom you are with and what you are doing.

Talk About It

People often think being aggressive makes them powerful. Do you agree?

Write About It

What style of communication do you use in school most of the time? Give an example.

What Is Body Language?

The words you use are an important part of communicating. But, words by themselves do not always give the whole message. Think back to Kris and James. Kris said nothing was wrong. But, by her actions, James knew something was wrong. He was reading Kris's body language. **Body language** is all the ways you use your body to communicate how you feel or what you think

Body language is very important. Suppose you go to a job interview. You tell the boss that you really want the job. But, while you are talking, you look at the floor. Then, you start biting your nails. Your body language is giving the boss a lot of information about you. It is saying that you are not confident. Or, maybe that you are not really interested.

The photo on the left is an example of positive body language. The photo on the right is an example of negative body language.

Practicing Positive Body Language

Your body language usually reflects how you feel about yourself. But, sometimes it works the other way. You can change how you feel just by changing your body language. Review the list below.

TIPS FOR ASSERTIVE BODY LANGUAGE

✓ Stand, walk, and sit tall. Practice good posture without being stiff.

✓ Look at someone for about four seconds at a time.

✓ Use your hands in a gentle, open way when you talk. Do not point; it suggests that you are blaming the other person. Crossing your arms shows you are not really open to talking.

✓ Use a friendly, natural smile, unless you are talking about something very serious. Giggling and frowning are not ways of being assertive.

✓ When you talk, take a look at your body language. Being aware of how you look is the first step toward change.

✓ Check Your Understanding

Suppose a friend owes you ten dollars and is late in paying it back. How could you ask for it in an assertive, but not aggressive, way? Write at least three possible ways on a separate sheet paper.

Tips for Spoken Communication

Do some people make you defensive right away? Do you feel like they are always blaming you for something? These people may not be communicating well. A good communicator:

- listens and does not blame.
- makes "I" statements instead of "you" statements.
- keeps a positive attitude.
- works as a member of a team to solve problems.

An Example of Good Communication

Think of this example. Roberta's mother is always telling her what kind of friends to have. If Roberta was a bad communicator, she might say to her mother, "You always make me feel like a baby. You never respect my decisions. You are ruining my life."

How do you think this would make Roberta's mother feel? It probably would make her feel as if Roberta was blaming her. She would probably say something angry to Roberta. Roberta would not get what she wants.

If Roberta was a good communicator, she would speak in a calm, thoughtful voice. She would say:

- "Mom, I want my choices to be respected."
- "I feel that I can handle making my own decisions about friends."

In this second example, Roberta speaks about her own feelings. She uses the word "I" rather than saying "you." When you say what you want, rather than telling someone else what they are doing, you are not blaming. You are stating your own feelings or thoughts.

Also, when you speak about what you want, rather than about what is wrong, you are being positive. Being positive almost always makes people more open to communicating.

Roberta might even ask her mother for help on the problem. "Mom, what can we both do so that I feel respected and you trust me?" A good communicator realizes that working together as a team gets better results than one person telling the other what to do.

Finally, a good communicator listens. Roberta should really listen to her mother's answer to the question. Sometimes, when you really listen to someone, they will do the same for you.

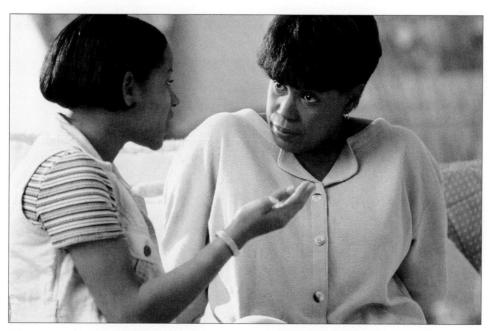

When you communicate well, the other person will be more likely to listen.

Healthy Relationships

Forming and keeping positive relationships is an important part of your mental and emotional health. You can have healthy relationships with your parents, your siblings, and other family members. You can also have healthy relationships with your peers. Good communication is necessary to maintain these relationships. The list below describes characteristics of a healthy relationship.

- Positive, open communication exists.
- Both people in the relationship feel attached and feel the relationship is important.
- Both people feel respected.
- The needs of both people are met.

Think about the relationships in your life. Do they fit the characteristics listed above? There will always be some amount of conflict in a relationship. **Conflicts** are disagreements between two people. Experiencing a conflict does not make the relationship unhealthy. But, if the characteristics listed above are rarely met, you may be in an unhealthy relationship.

Peer Relationships and Peer Pressure

As a teenager, your friends are probably very important to you. Some of your closest relationships may be with your peers. Having friends to share your interests and spend free time with is very healthy. But, it is important to choose your friends wisely. Your friends should listen to you, support you, and respect your opinions and values. If you have friends that do not do these things, they are not the best choices for a healthy friendship.

Talk About It

As you grow and develop, your friendships may change as well. This is a natural process. Are you friends with the same people now as you were five or ten years ago?

Health & Safety Tip

Dealing with conflicts early can help prevent the problem from becoming more serious. Conflicts that are left unresolved can often lead to arguments or even violence later.

Another sign of an unhealthy relationship is negative peer pressure. If you have friends who pressure you into doing something you do not want to do, they are not being good friends. In a positive relationship, you should never feel forced to think or act in a way that you do not want to. Instead of giving in to peer pressure, try suggesting a more positive activity. If that does not work, you should try to develop healthier friendships with different people.

Building Healthy Friendships

Sometimes making new friends can be hard. You may not feel comfortable around someone you have just met. But, as you get to know the person, you can decide if he or she would make a good friend. Below are three ways you can build healthy friendships:

1. **Find out about common interests and values.** Getting to know someone better means finding out their likes and dislikes. It also means finding out what you may have in common.

2. **Spend time together.** Spending time together doing something fun will help you to relax and enjoy each other's company. You can go shopping, play a sport together, or just talk.

3. **Learn to trust the other person.** This step is usually the hardest to do. It may only happen after you have gotten to know the person better. In order to develop trust, you will need to communicate in a healthy way. Talk openly and directly with the other person. Also, listen to him or her when they talk.

The list above describes how you can build healthy friendships. One thing that can hurt a friendship is gossip. **Gossip** is any negative comments made about another person without supporting evidence. You have probably heard plenty of examples of gossip. Gossip can be very hurtful. The best way to avoid gossip is to walk away or change the subject.

Remember
Peer pressure occurs when someone is forced to think or act in a certain way, just to fit in.

Write About It
Most people think of negative things when they think of peer pressure. However, peer pressure can also be positive. What are some ways your friends influence you in a positive way?

✓ Check Your Understanding

Write your answers in complete sentences.

1. Name three ways to communicate in a healthy way.

2. What is necessary in all relationships?

3. What are three characteristics of a good friend?

4. CRITICAL THINKING Why do you think gossip is unhealthy?

Think About It

HOW DO SCHOOL SOCIAL GROUPS AFFECT BEHAVIOR?

Schools are communities of learners. Like other communities, a school community can be broken down into smaller groups. These smaller groups are the people you socialize with. Research shows that the members of a social group usually share similar values and goals. They usually share common interests and enjoy the same activities.

Your friends may affect the decisions you make.

A recent study showed that a teen's social group greatly affects his or her decision to try risky behaviors. Teens whose closest friends smoked or drank alcohol were nine times more likely to mimic these behaviors than teens without friends who smoked or drank. The researchers concluded that the behavior of a teen's five closest friends was the most important factor in determining the teen's behavior.

Some people believe that many teens decide they want to try risky behaviors before joining a social group. They then become friends with peers who are likely to practice such behaviors. What do you think?

YOU DECIDE How do your values and goals compare with those of your five closest friends? Do you think you became friends because you already had these goals and values in place? Or did your goals and values change due to the influence of your friends?

LIFE SKILL
Resolving Conflicts

Most people try to avoid conflicts if they can. Sometimes they pretend a conflict does not exist. This approach is not very effective. Pretending a conflict does not exist can make it worse. Sometimes a problem seems worse the more you think about it.

It is better to try to resolve a conflict. Communication is a good way to resolve conflicts. But, you have to be careful to make sure that you communicate with respect. You do not want to say things that are hurtful. Here are some ways you can resolve a conflict:

Resolving a conflict between friends is better than pretending the conflict does not exist.

- Admit that the conflict exists.
- Talk calmly about the conflict.
- Be honest about your point of view and what you want to happen.
- Listen carefully to the other person and consider his or her feelings.
- Discuss different ways you could settle the conflict.
- Seek a solution that gives each person as much as possible. You and the other person may have to give something up to reach a compromise.

Sometimes people need help resolving conflicts. A mediator can help. A *mediator* is a person who is not involved in the conflict. The mediator tries to help find a solution that is acceptable to both sides.

Think about a conflict in the world today. Do you think the same skills that help solve personal conflicts can help solve larger conflicts? Give an example and explain why or why not.

Applying the Skill
Think about the last time you and a friend had a conflict. What was it about? How did you solve the conflict? Write about how you and your friend resolved or could have resolved the conflict.

Summary

Poor communication can lead to stress and anger. It can also lead to poor physical and mental health.

There are three main styles of communication. They are passive, aggressive, and assertive. The assertive communicator is the most healthy.

People communicate with their bodies. An assertive person shows confidence by standing tall and making eye contact.

A good communicator does not lay blame. He stays positive and asks for help in solving problems.

A healthy relationship is one in which both people feel respected and whose needs are met.

aggressive

assertive

body language

conflict

gossip

passive

Vocabulary Review

Complete the sentences with a word from the list.

1. A disagreement between two or more people is a _____.

2. To act in a pushy, dominating way is to be _____.

3. To be inactive, or unable to say what you think, want, or need is to be _____.

4. The ways you use your body to communicate how you feel or what you think is called _____.

5. Comments made about another person without supporting evidence are called _____.

6. To act in a positive, healthy way is to be _____.

Chapter Quiz

Write your answers on a separate sheet of paper. Use complete sentences.

1. How could poor communication lead to poor mental health?

2. What physical problems are caused by poor communication?

3. Name three benefits of being a good communicator.

4. How is communication inherited and learned?

5. How might a passive person act?

6. How might an aggressive person act?

7. How might an assertive person act?

8. What are three signs of a healthy relationship?

9. Describe how you can build a healthy friendship.

CRITICAL THINKING

10. If someone pressures you into doing something you do not want to do, are they being a good friend? Why or why not?

11. Suppose you find out a friend has been gossiping about you. Using good communication skills, write what you would say to this friend.

HEALTH LINKS.

Go to www.scilinks.org/health. Enter the code **PMH390** to research **communication skills**.

Online Health Project

Communication skills are something you will continue to improve for the rest of your life. These are skills that you will use at work, with your family and friends, and in your daily life. Find out more about communication skills. Make a poster that describes several communication skills. List an example of how each skill can be used to maintain positive relationships.

Cleaning up trash is just one way that volunteers keep communities attractive and healthy.

Learning Objectives

- Describe what makes a healthy community.
- Explain four ways to help make a community healthier.
- Explain how pollution affects your health.
- List the jobs public agencies do.
- **LIFE SKILL:** Describe how to access information about indoor air pollution.

Chapter 21 ▷ A Healthy Community

Words to Know

community	an area where people live and work together
environment	all of the things that surround you
pollution	harmful materials that damage the air, water, or soil
litter	trash or garbage in places they do not belong
graffiti	drawings or writings that are placed on walls or structures without permission
public health	the health of a large group of people such as a community or nation
advocacy group	a group of people that works to improve the health of others

What Makes a Community Healthy?

A **community** is an area where people live and work together. Most often, people think of their city as a community. Your neighborhood is also a community. Your school is also a community.

It is easy to identify things in your community that are not healthy. Drugs, crime, and joblessness are some examples. The list below describes a few things that make a community healthy.

- healthy individuals
- safe streets, schools, and parks
- clean air, water, and streets
- good care for the sick, elderly, and needy

People who live in the community determine what it will be like. This chapter looks at some ways you can help make your community healthy.

Talk About It

What are the two biggest health problems in your community? What are two healthy parts of your community?

Keep Yourself and Others Healthy

By keeping yourself healthy, you are helping your community to stay healthy. You can stay healthy by eating right, getting enough exercise, avoiding drugs and alcohol, and reducing your risk of disease. Practicing good hygiene can help prevent disease from spreading. Remember, if you get an STD, it is your responsibility to keep it from spreading.

Keep the Community Safe

Safety is a big part of staying healthy. This goes for your own health and the health of your community. What steps can you take to keep your community safe? Here are some ideas.

Write About It

Many communities have a neighborhood watch program that helps to prevent crimes. Find out if your community has one and how you can get involved.

Promote Safety in Your Community

- **Drive carefully.** Obeying traffic laws keeps you and the people in your community safe.

- **Do not get involved in crime.** Staying away from crime and criminals will keep you and those around you safe.

- **Report crimes or suspicious behavior.** Reporting crimes to the authorities will keep your community safe.

- **Help prevent violence.** You can prevent violence by resolving conflicts peacefully, avoiding violent people, and staying away from guns.

- **Stay away from drugs.** By not using drugs, you are promoting your own health and the health of your community. You are also avoiding the violence that sometimes is a part of drug use.

Keep the Community Clean

A clean community is a healthier community. You can help keep your community clean by finding out about the environment. Your **environment** is made up of all the things that surround you. The air, water, and soil are part of the environment. Living things such as grass and trees are also part of the environment.

Pollution and Your Health

Most environments have been affected by pollution. **Pollution** is damage to water, air, soil, or other parts of the environment by harmful materials. Pollution has negative effects on the environment and your health. When you drink polluted water, you can get sick. You can also get sick when you breathe polluted air. Harmful substances can pollute lakes, rivers, and oceans. If you eat fish or shellfish that come from polluted waters, you can get very sick.

If you want to help keep your community healthy, use trash cans.

Reducing Pollution in Your Community

The most obvious example of pollution in your community is litter. **Litter** is trash or garbage in places they do not belong. Candy wrappers, unfinished food, cans, and bottles are all examples of litter. Litter is a health threat in a number of ways. Rats, flies, and roaches gather where there is litter. These animals may carry bacteria and other pathogens around the community. Litter is also ugly.

Another form of pollution is called graffiti. **Graffiti** is any drawings or writings that are placed on walls or structures without permission. Graffiti may be scratched, painted, or sprayed onto buildings, bridges, and sidewalks. Some people think it is fun. But, graffiti damages private and public property. It is illegal.

Living in a community that is covered in litter and graffiti is not good for people's emotional health. It is depressing. A clean, litter-free and graffiti-free community is one that people can be proud of.

Ozone is a substance that can be harmful to your health if there is too much of it in the air you breathe. News programs often report ozone levels as part of their weather forecast.

There are also less noticeable types of pollution than litter—air and water pollution. The best way to reduce air and water pollution in your community is by learning about where the pollution comes from. Then you can find out what can be done to reduce the pollution.

You can reduce air pollution on your own by carpooling, taking a bus, or using a car that pollutes less. You can also find out if the factories in your community are following guidelines for reducing pollution. You can work together with other members of the community to keep your air and water clean.

Think About It

SHOULD THERE BE MORE ZERO-EMISSION CARS?

In 1990, California lawmakers passed a law to reduce air pollution in their state. The law required that a certain percentage of all cars sold in California be *zero-emission vehicles (ZEV)*. This law put pressure on carmakers to produce "clean" cars—or cars that do not release harmful pollutants into the air.

Fuel-cell cars produce no harmful pollutants.

Carmakers have designed a car that runs on hydrogen. They call this car a "fuel-cell" car. Fuel cells inside the car use hydrogen and oxygen to produce electricity. The electricity flows into a motor that supplies the power needed to move the car. No harmful pollutants are released into the air. Because hydrogen is the most plentiful element, car owners do not have to worry about using too much of a natural resource. As carmakers improve the design and cost of these vehicles, fuel-cell cars may become widely used.

YOU DECIDE Fuel-cell cars have to be refueled just like traditional cars. But, hydrogen-filling stations do not exist. Do you think lawmakers should pass another law requiring fuel companies to create hydrogen filling stations? Give reasons for your answer.

Volunteer in Your Community

No community is perfect. But, you can do your part to make it as good as it can be. You can help by volunteering your time in a community agency.

Talk About It

What job do you hope to have when you get older? Where could you volunteer to get experience for that job?

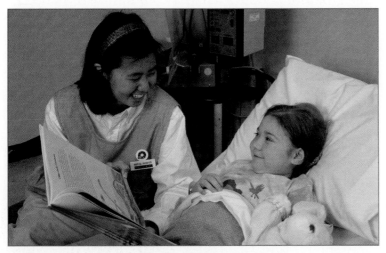

Volunteering helps members of the community.

Volunteering is good for the community because it helps other members of the community in many ways. It is good for you because you can learn skills that will help you get a job later in life. You can learn about careers you might want to explore when you get older. You can build confidence and "people skills."

Where can you volunteer? Hospitals, schools, nursing homes, drug programs, libraries, and churches can all use help. The American Red Cross, the American Cancer Society, and the Muscular Dystrophy Association are a few agencies that can always use help, too. Look them up in your local phone book or on the Internet.

✓ Check Your Understanding

Describe three things you can do to keep your community healthy. Be specific. Write your answer on a separate sheet of paper.

Public Health Agencies Are There to Help

Fortunately, you are not the only one keeping your community healthy. Tax dollars help pay for public health agencies. **Public health** means the health of a large group of people such as a community or nation.

There are many different types of public health agencies. Some of these agencies work to keep the environment clean. Some get help for sick people. Other public health agencies help educate people on ways to stay healthy. There are public health agencies that focus on small communities, such as a town or city. There also are public health agencies that focus on greater numbers of people, such as for a state, country, or even the world.

Examples of Public Health Agencies

Some public health agencies make sure restaurants and grocery stores keep foods fresh and clean. They see that food is handled correctly, and that restaurants are free of insects. Health agencies cannot always do all they should do with restaurants, though. If you see something wrong with your food, let a local agency know about the problem.

Public Health agencies may inspect restaurants like this one.

Public health agencies make sure that public water is germ-free. They help plan for new water sources. They try to clean up polluted water. They make sure that businesses and people do not pollute the water. If there is a problem with the drinking water, public health agencies will inform the public about the danger.

Public health agencies make sure septic tanks work properly. A septic tank collects and breaks down wastes flushed from homes called sewage. Public agencies may inspect new septic tanks. They may also test how a septic system is working every few years.

Public health agencies are the first to help after a disaster. A disaster can be caused by fire, flood, hurricane, earthquake, or tornado. People in the community may get water, shelter, or food from the agencies.

The American Red Cross helps people who are victims of disasters and emergencies.

Public health agencies often keep records of births, deaths, marriages, and divorces. These records help communities keep track of the health of the individuals that live in the community.

Perhaps the most important job of public health workers is treating and preventing diseases. They accomplish this by helping with family planning, prenatal care, and drug education. They keep records and inform people about communicable diseases. They help people find mental health services. They teach people about nutrition and exercise. They may bring prepared meals to disabled people. They provide hundreds of services that help protect and promote the health of others.

Talk About It

Would you enjoy a career in health? Why or why not?

You Can Make a Difference

Sometimes the problems of our communities seem too big. But, big problems are often solved by the individual efforts of people. Working together with other members of your community is the key to success.

You can advocate for a healthier community in many ways. To *advocate* means to give your support to something. An **advocacy group** is a group of people that works together to help others. Different advocacy groups focus on different aspects of the community, such as the environment or helping children. By joining an advocacy group related to your interests, you will be able to enjoy yourself while helping others.

 Check Your Understanding

Write your answers in complete sentences.

1. What are public health agencies?

2. What does it mean to advocate for something?

3. **CRITICAL THINKING** How would your community be affected if there were no public health agencies?

LIFE SKILL
Accessing Information About Indoor Air Pollution

Clean air is important for a healthy community. When people talk about clean air, they usually refer to the air outside. But, Americans spend about 90 percent of their time indoors. The quality of indoor air is also a major health concern. The list below shows some indoor air pollutants:

Indoor Air Pollutants	
• Cigarette smoke	• Carbon monoxide
• Pollen	• Mold spores
• Pet dander	• Cooking smoke
• Dust mites	• Fireplace smoke
• Lead paint	• Household chemicals

This image of a dust mite has been magnified 1,000 times.

Indoor air pollutants can have health effects. They can cause itchy eyes and runny noses. Other problems can be more serious, such as asthma and cancer. Breathing large amounts of carbon monoxide can cause death. Some people are more sensitive to certain indoor air pollutants than others.

It is easy to reduce air pollution. Many agencies provide information about how to do it. Here are some sources of information:

- U.S. Environmental Protection Agency http://www.epa.gov/
- American Lung Association http://www.lungusa.org/
- Local or state health departments

To conserve energy, newer buildings are more "tightly" constructed than older buildings. This keeps heat from escaping during cold weather. Do you think this makes indoor air pollution better or worse? Why?

Applying the Skill
Use sources of health information to find out how to reduce indoor air pollution at home. Create a poster to communicate this information to people in your school and community.

Summary

Everyone has a responsibility to keep their community healthy.

There are ways you can help to keep your community healthy. Some examples are keeping yourself and others healthy, keeping your community safe, keeping your community clean, and volunteering.

Public health agencies do many jobs. They keep the air and water clean, they keep health records, and they help keep food safe. They may help after a disaster. They also have a big part in preventing and curing diseases.

advocacy group

community

environment

litter

pollution

public health

Vocabulary Quiz

Match the definitions with a term from the list.

1. All of the things that surround you make up the _____.

2. The health of a large group of people such as a community or nation is called _____.

3. Harmful materials that damage the air, water or soil are called _____.

4. An area where people live and work together is called a _____.

5. Trash or garbage is called _____.

6. A group of people that work to improve the health of others is called an _____.

Chapter Quiz

Write your answers on a separate sheet of paper. Use complete sentences.

1. What is a community?
2. Name two things that make a community unhealthy.
3. What can you do to keep your community healthy?
4. How can pollution hurt your health?
5. How can you help prevent air pollution?
6. How can spreading STDs hurt your community?
7. What are two ways you can keep your community safe?
8. What are the benefits of volunteering?
9. How do health agencies help keep your food clean?
10. What other jobs do public health agencies do?

CRITICAL THINKING

11. How can litter and other forms of pollution hurt your mental health?
12. Think about your role in the community. Do you do enough to keep your community healthy? If so, give an example. If not, write one thing you could do.

Online Health Project

The health of different groups of people or nations is often closely related to the drinking water standards in that area. Find out how the quality and availability of water and human health issues are related. Describe specific examples from different parts of the world. Summarize your findings in a report.

> **HEALTH LINKS**
> Go to www.scilinks.org/health.
> Enter the code **PMH400** to research **drinking water standards**.

Unit 6 **Review**

Comprehension Check

On a separate sheet of paper, write how each of the things below will benefit your health.

1. being an assertive communicator
2. showing respect to the other person in a relationship
3. picking up litter in your neighborhood
4. volunteering in your community

Analyzing Cause and Effect

Write a sentence or two explaining what might cause the following. Use the information you learned in Chapters 20–21 for help.

5. A disagreement between two friends is settled in a good way.
6. A person does not get a job because he is slouching in the interview.
7. Disease breaks out in a building full of rats and litter.
8. A restaurant is closed by the local health agency.

Writing an Essay

Write an essay to answer each of the following questions.

9. What are the four characteristics of a healthy relationship? Give examples for each.
10. How can peer pressure make you unhealthy?
11. Why is a healthy community important to your own health?
12. What are some ways to keep your community safe?

Managing Your Health

Write a few paragraphs about how your communication skills affect your health. Include examples of poor and good communication.

Appendix A: Food and Calories

Food Group	Kind of Food	Size of Serving	Food Energy in Calories*
Meat, Poultry, Fish, Dry Beans, Eggs, and Nuts Group	Bacon	1 slice	150
	Beef	3 ounces	230
	Chicken	3 ounces	156
	Egg	1 large	90
	Fish	3 ounces	150
	Ham	3 ounces	187
	Hot dog	1	150
	Lunch meat	1 slice	185
	Peanut butter	1 tablespoon	90
Vegetable Group	Baked potato	1 medium	95
	Vegetables	1/2 cup	145
Fruit Group	Fruit (canned)	1/2 cup	160
	Fruit (fresh)	1/2 cup	100
Milk, Yogurt, and Cheese Group	Cottage cheese	1 ounce	25
	Other cheese	1 ounce	110
	Ice cream	1 serving	170
	Skim milk	1 cup	85
	Whole milk	1 cup	160
	Yogurt	1 cup	155
Bread, Cereal, Rice, and Pasta Group	Biscuit	1	130
	Bread	2 slices	120
	Cereal	3/4 cup	170
	Grits	1 cup	120
	Macaroni	1/2 cup	180
	Noodles	1/2 cup	180
	Rice	1/2 cup	100
	Roll, plain	1	85
	Soda crackers	2	45
	Spaghetti	1/2 cup	180
Fats, Oils, and Sweets Group	Butter	1 pat	50
	Cake with icing	1 slice	320
	Doughnut	1	135
	Fudge	1 piece	115
	Jelly	1 tablespoon	150
	Margarine	1 pat	50
	Mayonnaise	1 teaspoon	40
	Milk chocolate	1 bar	400
	Pie	1 piece	300
	Potato chips	10	100
	Salad dressing	1 tablespoon	150
	Sugar	1 tablespoon	55
	Sweet roll	1	100

* The Calories given for foods are averages.

Appendix B: Some Important Vitamins and Minerals

Vitamin	Source	Body Function
A	Leafy green and yellow vegetables, egg yolk, milk, liver, butter, margarine	Good eyesight; healthy skin and hair; growth
B_1	Whole grains, yeast, milk, green vegetables, egg yolk, liver, fish, soybeans, peas	Strong heart, nerves, and muscles; growth; respiration
B_2	Lean meat, wheat germ, yeast, milk, cheese, eggs, liver, bread, leafy green vegetables	Healthy skin; growth; good eyesight; reproduction
B_{12}	Liver, lean meat, milk, fresh fish, egg yolk, shellfish	Helps make blood; helps nervous system
C	Oranges, grapefruit, lemons, limes, berries, vegetables, tomatoes	Healthy bones; strong blood vessels; helps heal wounds
D	Egg yolk, milk, fresh fish	Strong teeth and bones; growth
E	Leafy green vegetables, wheat germ, oils	Healthy skin; prevents cell damage
K	Vegetables, soybeans	Helps blood clotting

Mineral	Source	Body Function
Calcium	Milk, vegetables, meats, dried fruits, whole-grain cereals	Healthy bones and teeth; helps blood clotting
Iodine	Saltwater fish, shellfish, iodized salt	Regulates use of energy in cells
Iron	Liver, meats, eggs, nuts, dried fruits, leafy green vegetables	Forms red blood cells
Magnesium	Milk, meats, whole grain cereals, peas, beans, nuts, vegetables	Normal muscle and nerve action; regulates body temperature; helps build strong bones
Phosphorus	Milk, meat, fish, poultry, nuts, vegetables, whole-grain cereals	Forms bone and teeth; nerve and muscle function; produces energy
Potassium and sodium	Most foods, table salt (sodium)	Blood and cell functions; help maintain balance of fluids in tissues

Appendix C: Exercise

Activity	Calories Used Per Hour	Health Benefits
Bicycling: 6 mph	240–300	Good for heart; strengthens muscles
11 mph	420–480	Builds endurance; strengthens muscles; excellent for heart
Cleaning house	240–300	A good workout if cleaning for at least 20 minutes
Floor exercises (such as sit-ups and push-ups)	300–360	Good for muscles and body tone
Jogging (5 mph)	480–600	Very good for heart and muscles; builds endurance
Running (6 or more mph)	660 or more	Excellent for heart and muscles; builds endurance
Sitting	72–84	No physical benefits; can relieve stress
Swimming	Depends on style and speed	Good for heart and muscles
Walking: 2 mph	120–150	Tones muscles
3 mph	240–300	Good for heart and tones muscles
4 mph	360–420	Good for heart and tones muscles

Source: Adapted from "Beyond Diet . . . Exercise Your Way to Fitness and Health" by Lenore R. Zohman, M.D., copyright CPC International, Inc.

Appendix D: Diseases

Disease	Signs	Health Effects if Left Untreated
AIDS (Aquired Immune Deficiency Syndrome)	First signs may be weakness, swollen glands, diarrhea	No cure at this time; may lead to death
Asthma	Shortness of breath, wheezing, coughing, tightness in chest	Continued asthmatic episodes
Cancer	Changes in skin, unusual bleeding, a sore or cut that does not heal, coughing that does not go away, difficulty swallowing, an unusual lump in the breast or testes	No cure for most forms of cancer, but may be put into remission through radiation, chemotherapy, or surgery; may lead to death
Cardiovascular disease	Pain in the center of the chest, irregular heartbeat, hypertension	Hardened or clogged arteries, heart attack, stroke, death
Diabetes	Urinating a lot and being very thirsty (Type I and Type II diabetes); appetite loss and vomiting (Type I); hunger, some weight loss, tiredness (Type II diabetes)	Heart disease, stroke, possible blindness or loss of appendages
Epilepsy	Seizures, which may include shaking or sudden movements, suddenly falling down, or loss of consciousness for a short time	Continued seizures, brain damage may result from a blow to the head during a seizure
Gonorrhea (clap, drip)	In men, drip from penis; pain and burning when urinating. In women, there are often no signs.	Blindness, sterility
Hemophilia	Tendency to bleed a lot; large bruises after minor injuries; swelling in the knees, ankles, elbows	May lead to anemia, a condition in which red blood cells are reduced, and even death
Herpes	Sores on the sex organ or mouth	Some people become tired and feel as if they have the flu; others feel nothing. Pregnant women with herpes must tell their doctors; the disease can hurt unborn children.
Syphilis	Sores on the mouth or sex organs	Blindness, insanity, death
Venereal warts	Warts on or near the sex organs	No cause for worry
Yeast infection	Sex organs itch	Itching, vaginal discharge may continue

Appendix E: First Aid

You can avoid most accidents by following the tips given in Chapter 8 of this book. However, accidents can still happen, almost anywhere. Therefore, it is a good idea to be prepared for them.

First aid is emergency medical care given to someone who has just been injured or become very sick. Different emergencies need different kinds of first aid. Here are some rules to follow for any emergency.

- Stay calm so that you can think clearly.
- Call 911 or 0 to reach Emergency Medical Services (EMS), or ask someone nearby to call.
- Do not move the person unless he or she is drowning or in danger of being burned in a fire.
- Check for signs of breathing and a pulse.
- Check for any emergency medical ID tag or bracelet.
- Never do more than you know how to do. Find help if the injury seems serious, if you do not see breathing or cannot find a pulse, if bleeding continues, or if a pain is very sharp.

The chart on page 309 provides some basic guidelines. However, the best way to learn first aid is to take a course that teaches first-aid techniques such as cardiopulmonary resuscitation (CPR) and the Heimlich maneuver, which is a way to stop someone from choking on an object stuck in the windpipe. First-aid courses are available through libraries, fire departments, hospitals, the Red Cross, and other organizations in your community.

Appendix E: First Aid (*continued*)

Type of Injury	First Aid Treatment
Bruising	• Apply ice or cold compresses to reduce swelling.
Choking	• If the person cannot talk, cough, or breathe, perform the Heimlich maneuver.
Cuts	• Cover and apply direct pressure to the wounds. • Use sterile or clean bandages.
Fainting	• Make sure the person is lying flat with feet up. • Use cool water on the face. • Loosen any tight clothing.
Insect stings	• Scrape the stinger away; do not pull it straight out. • Rinse the skin with cool water. • If the person is allergic, take to an emergency room immediately.
Most burns	• Apply cool water. • Place clean, dry bandages loosely over the burns. • Do not apply ointments or sprays. • See a doctor if burns look bad.
Nosebleeds	• Make sure the person is leaning forward while breathing through the mouth. • Use a wet cloth and keep pressure on the nostril for 10 minutes. • Do not block the airway.
Objects in an eye	• Flush the eye with water, move from the inner to the outer eye corners.
Poisoning	• Find the container the poison came in and read the label instructions. • Contact your local poison control center.
Sprains	• Raise the sprained body part. • Help reduce swelling with ice or a cold pack.
Stopped heartbeat	• Use cardiopulmonary resuscitation (CPR). Call 911.

Glossary

A

abstain to choose not to drink (p. 220)

abstinence deciding not to have sex and sticking with your decision (pp. 105, 270)

acid a slang term for *LSD* (p. 234)

acne any type of clogged skin pore (p. 134)

acquired immunity an ability to resist a certain disease that develops over time (p. 48)

action plan a list of steps that will help you reach a goal (p. SHL6)

acupuncture a type of alternative medicine in which a needle is placed through the skin of the sick person at certain points (p. 204)

addiction a need or habit for a substance (p. 219)

addictive habit forming (p. 231)

adolescence the teenage, or young-adult, years (p. 253)

adrenaline a hormone that gives the body extra energy and strength (p. 201)

advocacy group a group of people that works to improve the health of others (p. 298)

advocate to give your support to something (p. 298)

aggressive to act in a pushy or dominating way (p. 279)

AIDS Acquired Immune Deficiency Syndrome; a disorder that affects the body's ability to fight disease (p. 103)

alcoholic a person who is addicted to alcohol (p. 219)

alcoholism a disease that causes a person to be addicted to alcohol (p. 219)

alveoli tiny air sacs in the lungs that exchange gases; singular is *alveolus* (p. 51)

Alzheimer's disease an illness that weakens and kills brain cells, and destroys the victim's powers of memory and reasoning (p. 162)

amino acid the smallest part of a protein (p. 143)

amniotic sac a pouchlike structure that holds a developing embryo or fetus (p. 79)

amphetamine a drug that stimulates or speeds up the nervous system and heart rate (p. 232)

angel dust a slang term for PCP (p. 235)

anorexia a disorder that causes a person not to eat (p. 195)

antibiotic a medicine that kills harmful bacteria (p. 91)

antibody a molecule that attaches to a specific pathogen (pp. 48, 95)

antiperspirant a product that helps to control perspiration (p. 134)

antiseptic a substance that kills bacteria and other germs (p. 90)

anus an opening through which waste matter in the large intestine passes out of the body (p. 64)

anxiety a deep fear or worry that something bad is going to happen (p. 188)

appendicitis an infection of the appendix (p. 58)

arteriogram a type of screening test that shows blocked blood vessels (p. 111)

artery a blood vessel that carries blood away from the heart (p. 41)

arthritis a condition that causes stiff and swollen joints (p. 10)

assertive to act in a positive, healthy way (p. 279)

asthma a type of lung disease that affects breathing (p. 109)

atria the two top chambers of the heart (p. 40)

automatic impulse   an involuntary impulse that controls how your body's organs function (p. 26)

axon   the fiber on neurons that carries messages away from the cell (p. 25)

B

bacteria   types of one-celled organism that can cause disease; singular is *bacterium* (p. 93)

balanced diet   a diet that contains foods from the four food groups (p. 142)

ball-and-socket joint   a joint that allows bones to move in several directions; the shoulder, for example (p. 7)

barbiturate   a drug that slows down the nervous system (p. 231)

benign tumor   a tumor that does not spread to other organs (p. 106)

biceps   the muscle in the front of the upper arm (p. 13)

bile   a green liquid made by the liver that helps digest fats (p. 60)

binge   to take in food or drink in extremely large amounts at one time (p. 195)

blood alcohol level (BAL)   a measurement of how much alcohol is in the blood (p. 217)

body language   the ways you use your body to communicate how you feel or what you think (p. 280)

body system   a group of organs that work together (p. 3)

bone cement fixation   a method used for joint replacement in which an artificial joint is cemented between two bones (p. 11)

brainstem   the part that connects the rest of the brain to the spinal cord (p. 23)

brain tumor   abnormal growth of nerve cells in the brain (p. 161)

bronchi   small tubes in the trachea that air moves through into the lungs; singular is *bronchus* (p. 50)

bronchitis   an irritation or infection of the airways (bronchi) that causes a coughing condition (p. 223)

bulimia   a disorder that causes a person to overeat and then try to get rid of the food just eaten (p. 195)

C

Calorie   a measure of the heat energy found in foods (p. 152)

cancer   the uncontrolled abnormal growth of cells (p. 106)

Cannabis sativa   the scientific name for marijuana (p. 235)

capillary   a tiny blood vessel that connects arteries to veins (p. 41)

carbohydrates   a substance found in foods that provides energy (p. 143)

carbolic acid   one of the first-known germ killers (p. 90)

carbon monoxide (CO)   a poisonous gas that comes from burning materials (p. 117)

carcinogen   a cancer-causing substance (p. 106)

cardiac muscle   the muscle tissue that makes up the heart (p. 13)

cardiovascular disease   a disorder that affects the heart and blood vessels (p. 108)

cartilage   a tough, spongy tissue (p. 5)

cataracts   a condition that occurs when the lens becomes blurry and limits vision (p. 32)

cells   the basic units of structure and function in living things (p. 3)

cerebellum   the part of the brain that controls balance and coordination (pp. 23, 160)

cerebral palsy   a disorder of the nervous system caused by brain damage resulting in less-than-normal control over muscles (p. 27)

cerebrum   the part of the brain that controls voluntary muscle movements, thinking, learning, memory, speech, and the senses (pp. 22, 160)

cervix the opening to the uterus (p. 76)

chamber open space in an organ such as one of the four chambers of the heart (p. 40)

chemical digestion a chemical reaction that breaks down food during digestion (p. 58)

chemotherapy the use of drugs to kill cancer cells (p. 107)

chlamydia a sexually transmitted disease that affects the vagina in females and the urethra in males (p. 105)

cholesterol a fatlike substance needed by the body in small amounts (p. 44)

chorea an infectious disease that attacks the brain (p. 161)

cilia the hairlike structures lining the lungs and airways (p. 94)

circulatory system the organ system that moves blood throughout the body (p. 40)

cirrhosis of the liver an often fatal disease that causes the liver to stop functioning (p. 219)

closed fracture a broken bone that stays in place (p. 8)

co-insurance the amount you pay after health insurance pays its part of a bill (p. SHL11)

coke a slang term for *cocaine* (p. 233)

colitis an inflammation of the colon, or large intestine (p. 62)

colon the large intestine (p. 61)

colonoscopy an examination that uses a thin tube with a small camera to look inside the colon (p. 63)

communicable disease a disease that is contagious and can be passed from one person to another (p. 92)

community an area where people live and work together (p. 291)

compact bone the hardest part of a bone made up of living bone cells (p. 5)

compensation becoming strong in one area to make up for a weakness in another area (p. 177)

compulsive doing something over and over again (p. 188)

conflict a disagreement between two or more people (p. 284)

constipation a condition that occurs when wastes do not move through the large intestine on a regular basis (p. 61)

contagious disease a disease that can be passed from one person to another (p. 92)

contractions strong muscle movements in the mother's uterus that force the baby out of her body (p. 81)

co-payment (co-pay) a fixed dollar amount for a healthcare service, such as a doctor's visit (p. SHL11)

cope to face or deal with a problem or responsibility (p. 205)

cornea the clear, curved covering on the outer surface of the eye (p. 30)

coronary artery a blood vessel that brings nutrients and oxygen to the muscles of the heart; if blocked, can cause a heart attack (p. 45)

cortisone a medicine that reduces swelling; used for arthritis (p. 10)

crank a slang term for *amphetamines* (p. 232)

crisis an extremely difficult situation (p. 182)

custody a legal arrangement that says who is responsible for the care and well-being of a child (p. 248)

cystic fibrosis an inherited disease that causes problems in digestion and breathing (p. 110)

cystitis a bladder infection (p. 66)

D

dandruff dead skin flaking off the scalp (p. 136)

daydreaming a defense mechanism that allows a person to escape through imagination (p. 178)

decision-making model   a well thought-out process with seven steps that can help you make healthy choices (p. SHL5)

deductible   a yearly amount you must spend before health insurance pays anything (p. SHL11)

defense mechanism   a way of coping with emotions (p. 176)

deficiency   a lack of a nutrient (p. 148)

dehydration   the drying out of your body cells due to lack of water (p. 67)

dendrite   the fiber on neurons that carries messages into the cell (p. 25)

denial   not facing emotions or problems (p. 179)

dentin   the middle layer of a tooth (p. 132)

deodorant   a product that helps control body odors (p. 134)

depressant   a drug that slows down the nervous system (p. 216)

depression   a state of mind that causes a person to withdraw and feel sad for long periods of time (p. 192)

dermatologist   a doctor who specializes in treating skin problems (p. 135)

diabetes   a disease that affects the way the body changes food into energy (p. 109)

dialysis   an artificial way to clean the blood by using a machine (p. 67)

diaphragm   a large skeletal muscle that pulls the bottom part of the chest cavity downward during breathing (p. 50)

digestion   the process by which the body breaks down food into nutrients that can be absorbed by cells (p. 57)

displaced aggression   taking anger out on a person or thing that did not cause the anger (p. 176)

divorce   a legal end to a marriage (p. 248)

downers   a slang term for barbiturates (p. 231)

driving under the influence (DUI)   driving a car while under the influence of alcohol above the legal limit; this is a serious crime (p. 218)

driving while intoxicated (DWI)   driving a car while under the influence of alcohol above the legal limit; this is a serious crime (p. 218)

drug   a chemical substance that affects the body systems (p. 214)

drug-free school zone   areas around schools in which penalties for using or selling drugs are much higher than in other areas (p. 237)

dysfunctional   something that does not function correctly (p. 246)

E

eardrum   a thin layer of tissue in the ear that transfers sound vibrations from the outer ear to the middle and inner ear (p. 30)

egg cell   a female reproductive cell (p. 76)

electrical shock   a response to electricity passing through the body (p. 119)

embryo   an organism that is developing from a fertilized egg (p. 78)

emotional health   a person's ability to cope with his or her emotions (pp. SHL2, 174)

emotions   feelings such as fear, anger, sadness, happiness, and love (p. 163)

emphysema   a serious disease of the respiratory system (pp. 52, 222)

enamel   the outer covering of a tooth (pp. 59, 132)

endocrine gland   an organ that produces chemical substances called hormones that are released into the bloodstream (p. 33)

endurance   the measure of how long a person can continue an exercise (p. 149)

environment   all the things that surround you (pp. SHL3, 293)

enzyme   a substance that helps to change chemical reaction rates in the body (p. 58)

epidemic a widespread occurrence of a certain disease (p. 92)

epiglottis a small piece of tissue at the top of the windpipe that closes during swallowing (p. 50)

epilepsy a brain disorder that causes a person to have seizures (pp. 28, 109)

esophagus a tube behind the windpipe that carries food from the mouth to the stomach (p. 59)

estrogen a female sex hormone (p. 76)

excretory system the organ system that removes wastes from the body (p. 64)

extended family a family made up of a nuclear family and other relatives, such as grandparents, aunts, uncles, and cousins (p. 245)

F

Fallopian tube a tube that leads from an ovary to the uterus (p. 76)

family a group of people that is related by genetics, marriage, or legal action (p. 245)

fertilization the joining of egg and sperm cells (p. 78)

fetal alcohol syndrome a disorder in babies born to mothers who drank alcohol while they were pregnant (p. 80)

fetus the term used to describe an embryo after 8 weeks of development in the uterus (p. 79)

fiber a form of carbohydrate that cannot be digested (p. 144)

fire extinguisher a metal device filled with water, chemicals, and gases that is used to put out fires (p. 117)

Food Guide Pyramid a diagram that shows the kinds of foods and numbers of servings of each that an average person should eat each day (p. 145)

fracture a break or crack in a bone (p. 8)

G

gallbladder an organ that stores bile (p. 60)

gay a term for a male or female homosexual (p. 266)

genetic engineering using laboratory techniques to change an organism's genetic material, or genes (p. 49)

genital herpes a sexually transmitted disease that causes blister-like sores in the genital area (p. 105)

genitals the external reproductive organs (p. 74)

germ theory the theory that germs, or microscopic organisms, cause disease (p. 90)

gland an organ that makes chemical substances that are used or released by the body (p. 25)

glaucoma an eye disorder that can cause blindness (p. 32)

gliding joints joints that allow bones to move in many directions, such as in the wrist (p. 7)

glucose a simple sugar formed when the digestive system breaks down starches and sugars (p. 143)

goal something you want to do or achieve; something that gives you direction (p. SHL6)

gonorrhea a sexually transmitted disease that affects the lining of the reproductive organs (p. 105)

gossip comments made about another person without supporting evidence (p. 285)

graffiti drawings or writings that are placed on walls and structures without permission (p. 291)

group health insurance insurance purchased through an employer (p. SHL11)

growth spurt a time when a person grows more quickly than at other times (p. 254)

guardian a person who cares for a child but is not the parent (p. 248)

H

halitosis bad breath (p. 133)

hallucinogen a drug that causes a person to see, hear, and feel things that are not real (p. 234)

hazard any unsafe condition that can cause harm to people (p. 119)

hazardous substance a poisonous chemical that can do harm or kill if swallowed or touched (p. 118)

health insurance a plan that helps you pay doctor and hospital bills (p. SHL11)

hemoglobin a protein in red blood cells that contains the mineral iron (p. 42)

hemophilia an inherited disease that affects the way blood cells clot (p. 110)

heredity the passing of certain traits from parents to their children (p. SHL3)

heterosexual a person who is attracted to those of the opposite sex (p. 266)

hinge joint a joint that allows a part of your body, such as an elbow, to move in only two directions—back and forth (p. 7)

HIV Human Immunodeficiency Virus; the virus that causes AIDS (p. 103)

homosexual a person who is attracted to those of the same sex (p. 266)

hormone a type of chemical messenger in the body that is produced by endocrine glands (p. 33)

horse a slang term for *heroin* (p. 234)

hygiene taking actions to keep yourself clean and healthy (p. 132)

hypertension high blood pressure (p. 44)

hypnotherapy a type of alternative medicine used to treat pain and stress in which a hypnotist focuses the patient's attention on the injured body part (p. 204)

I

ice a slang term for *amphetamines* (p. 232)

immovable joint a joint that does not allow movement, such as in the skull (p. 7)

immunity the ability to resist a certain disease (p. 48)

immunization record a record that shows which vaccines a person has received (p. 214)

impulse a signal sent through the nervous system to different organs and systems (p. 26)

infertility the inability to have children naturally (p. 81)

inhalant a substance that is inhaled (p. 236)

inherited passed down from a previous generation in a person's genetic material (p. 110)

intoxicated having a blood alcohol level of about 0.08 to 0.1 (p. 218)

involuntary muscle a muscle that moves without your control (p. 13)

iris the colored part of the eye that controls the size of the pupil (p. 30)

J

joint a place in the body where two or more bones meet (p. 7)

junk a slang term for *heroin* (p. 234)

K

kidney failure a condition that occurs when one or both kidneys shut down (p. 67)

L

laxative a medicine used to treat constipation (p. 61)

lens the part of the eye that focuses light on the retina (p. 30)

lesbian a term for a female homosexual (p. 266)

leukemia a type of cancer that affects the bone marrow and blood (p. 106)

life preserver a device that helps a person to stay afloat in water (p. 120)

ligament a type of tissue that holds bones together at a joint (p. 7)

litter   trash or garbage in places they do not belong (p. 293)

liver   a large organ that produces bile (p. 60)

lymph   a clear fluid that contains special white blood cells (p. 47)

lymph node   an organ that filters lymph (p. 47)

lymphocyte   a special white blood cell that fights disease (p. 47)

M

malignant tumor   a tumor that is harmful, spreads to other parts of the body, and can damage important organs (p. 106)

malnutrition   a condition that occurs when a person does not get enough food and nutrients (p. 148)

mammogram   an x-ray obtained from a technique that can detect breast cancer (p. 111)

marrow   the soft tissue inside bones that makes blood cells (p. 6)

mechanical digestion   occurs when food is broken down into smaller pieces in the digestive system but not changed chemically (p. 58)

media   refers to newspapers, magazines, radio, television, movies, music, videos, and the Internet (p. 137)

mediator   a person who tries to help find a solution that is acceptable to both sides during a conflict (p. 287)

medicine   a drug that is used to prevent or cure a disease or medical problem (p. 214)

medulla   the part of the brain that controls involuntary functions (p. 23, 160)

meningitis   an infectious disease that attacks the brain and spinal cord (p. 161)

menopause   the permanent end of menstruation (p. 77)

menstruation   the monthly shedding of the lining of a woman's uterus (p. 77)

mental disability   a condition that interferes with a person's ability to think, learn, or speak (p. 161)

mental disorder   a condition that disturbs a person's emotions, thinking, and behavior (p. 188)

mental health   wellness of the mind (p. 163)

mental illness   a disorder that seriously affects a person's emotions, thinking, and behavior (162)

mental retardation   a mental disability that limits a person's ability to learn (p. 161)

metabolism   the process by which energy from food is used to carry out life functions (p. 152)

meth   a slang term for *amphetamines* (p. 232)

methedrine   one of the most powerful amphetamines (p. 232)

mineral   a nutrient found in foods that the body needs and that is found in living and nonliving things (p. 146)

mononucleosis   a contagious disease that affects the white blood cells (p. 102)

motor neuron   a nerve that carries messages away from the central nervous system to muscles and glands (p. 25)

multiple sclerosis   a disease of the nervous system that leaves nerves hardened and scarred (p. 27)

muscular dystrophy   a disease that causes muscles to waste away over time (p. 15)

N

narcotic   a drug that comes from the opium plant (p. 234)

nephron   a network of tubes or vessels in a kidney that filters waste from blood (p. 65)

nerves   threadlike tissues that carry messages to and from the brain (p. 25)

network   a group of selected doctors that participate in a particular health insurance plan (p. SHL11)

neuron a nerve cell (p. 25)

neurosis a mild type of mental disorder that may be treated with therapy (p. 188)

neurotic a person with neurosis (p. 188)

nicotine the addictive drug in tobacco (p. 222)

non-communicable disease a disease that is not contagious and cannot be spread by contact among people (p. 92)

nuclear family a family made up of two parents and one or more children (p. 245)

nutrient a chemical substance the body cannot make but needs to stay alive (p. 142)

O

obsessive having the same thoughts that repeat over and over again (p. 188)

obsessive-compulsive disorder (OCD) a pattern of repeated thoughts and behaviors that interferes with a person's life (p. 188)

open fracture a broken bone with torn muscles and torn skin (p. 8)

organs groups of tissues that work together (p. 3)

osteoporosis a condition that makes bones weak and brittle (p. 9)

ovarian cysts masses of abnormal tissue in the ovaries (p. 82)

ovaries the female organs that make egg cells and hormones (p. 76)

overall health made up of physical, emotional, and social health (p. SHL2)

over the counter medicine (OTC) medicine that can be bought without a doctor's prescription (pp. SHL10, 214)

ovulation the monthly release of an egg cell from an ovary (p. 76)

P

pancreas an organ that produces digestive enzymes and hormones (p. 60)

panic disorder another form of neurosis (p. 189)

paranoia an unreasonable fear that someone is trying to harm you (p. 189)

passive to be inactive or unable to say what you think, want, or need (p. 279)

pasteurization a method of controlling bacteria in food by heating the foods (p. 91)

pedestrian a person who is walking (p. 124)

peer pressure pressure from friends to act or think in a certain way (pp. 236, 258)

penicillin an antibiotic that kills germs; the first antibiotic to be discovered (p. 91)

penis the male organ that delivers sperm to the female reproductive system (p. 75)

personal health assessment a survey that asks questions about your health and your knowledge of health risks (p. SHL9)

personality the combination of many different traits that makes each person unique (p. 167)

perspiration excess water and salts released from sweat glands (p. 68)

pharmakons ancient Greek word for *drugs* (p. 230)

phobia an unreasonable fear of an object or event (pp. 175, 189)

physical health the wellness of the body (p. SHL2)

pimple a clogged pore that becomes infected (p. 134)

pituitary gland a small gland in the brain that controls the release of hormones (p. 255)

pivotal joint a joint that allows bones to move up and down or from side to side (p. 7)

placenta the organ through which nutrients, oxygen, and wastes pass between the mother and the embryo or fetus (p. 78)

plaque a sticky covering on the teeth caused by bacteria (p. 132)

plasma the liquid part of blood that carries nutrients and wastes throughout the circulatory system (p. 42)

platelet a piece of a cell found in blood that helps blood cells clump together (p. 42)

pollution harmful materials that damage the air, water, or soil (p. 293)

posture the way that you hold yourself up when sitting or standing (p. 130)

pot a slang term for *marijuana* (p. 235)

premium a yearly fee for health insurance (p. SHL11)

prenatal before birth (p. 27)

prenatal care care a baby receives while it is inside its mother (p. 79)

prenatal ultrasound a scanning technology that helps monitor the progress of babies in the uterus (p. 111)

prescription an order for medicine (p. SHL10)

prescription medicine medicine that must be ordered by a doctor (pp. SHL10, 214)

progesterone a hormone that prepares the uterus to receive a fertilized egg (p. 76)

projection seeing your emotions in another person (p. 178)

protein a substance found in foods that builds and repairs body cells (p. 142)

psychiatrist a medical doctor who specializes in treating serious types of mental disorders (p. 190)

psychoactive drug a drug that changes the way a person's brain functions (p. 216)

psychological dependence a person's emotional need for a drug (p. 231)

psychologist a caregiver trained in giving therapy to people with mild mental disorders (pp. 181, 190)

psychosis a serious mental disorder that must be treated by a doctor (p. 190)

psychotic one who withdraws into him or herself (p. 190)

puberty the time at which a person becomes sexually mature (p. 74)

public health the health of a large group of people such as a community or nation (p. 296)

pulled muscle small tears in a muscle (p. 17)

pulp the core or center of a tooth (p. 132)

pupil the opening in the eye that lets light in (p. 30)

pus a white or yellow pool of dead white blood cells and other tissues (p. 94)

R

radiation the use of radioactive energy to destroy cancer cells (p. 107)

rationalization using weak or false reasons to hide the true reason for bad behavior (p. 177)

rectum lower end of the large intestine (p. 64)

red blood cell a type of cell found in blood that carries oxygen to cells (p. 42)

reflex an involuntary response to outside stimuli (p. 26)

remission a condition in which symptoms of cancer have disappeared (p. 307)

retina the layer of sensory neurons at the back of the eye that detects light (p. 30)

Rh factor a substance on the surface of a blood cell that is either positive or negative (p. 43)

Rh negative not having the Rh substance on blood cells (p. 43)

Rh positive having the Rh substance on blood cells (p. 43)

risk behaviors choices that put your health or the health of others at risk (p. SHL5)

roots parts of a tooth that contain nerves and blood vessels (p. 59)

rush the feeling a person experiences when a drug reaches the brain (p. 232)

S

saliva a liquid in the mouth that helps digestion (p. 59)

saturated fats   fats that help raise the cholesterol level in the blood (p. 143)

schizophrenia   a kind of psychosis (p. 190)

secondhand smoke   smoke that is in the air surrounding smokers, which can be harmful (p. 224)

seizure   a sudden attack during which a person loses control of his or her body functions (p. 109)

self-image   how a person feels about himself or herself (p. 174)

semen   the mixture of fluids in which sperm leaves the body (p. 75)

sensory neuron   takes messages away from the sense organs to the brain and spinal cord (p. 25)

separation   agreement between a married couple to stay married but live apart (p. 247)

sewage   wastes flushed from homes into septic tanks or public sewage systems (p. 297)

sexuality   the state of being sexual; a person's sexual interests and issues (p. 265)

sexually transmitted disease (STD)   a contagious disease that spreads from person to person by sexual contact (pp. 103, 268)

sickle-cell anemia   an inherited disease that affects the shape and function of red blood cells (p. 110)

side effect   a reaction to medicine that is different from the reaction the medicine is supposed to cause (p. 215)

single-parent family   a family that has only one parent, either a mother or father, and one or more children (p. 245)

skeletal muscles   voluntary muscles attached to bones (p. 12)

sleep apnea   when a person stops breathing periodically during sleep (p. 131)

smack   a slang term for *heroin* (p. 234)

smoke detector   a device that warns people of fire or smoke by letting out a loud noise (p. 117)

smooth muscle   a muscle that is not controlled and is not attached to bones (p. 13)

social drinker   a person who has no more than a couple of drinks with friends now and then (p. 219)

social health   the ability to get along with others (p. SHL2)

somnoplasty   a technique used to remove extra tissue in the upper airways (p. 131)

Special K   a slang term for *amphetamines* (p. 231)

speed   a slang term for *amphetamines* (p. 232)

sperm cell   a male reproductive cell (p. 75)

sperm ducts   thin tubes in which sperm travels from the testes (p. 75)

spinal cord   a ropelike structure that is made up of many nerve cells (p. 23)

spongy bone   softer parts of bone (p. 5)

starches   complex carbohydrates (p. 143)

sterilize   to make free of germs, usually by hot water, steam, or chemicals (p. 90)

stimulant   a type of drug that speeds up the nervous system and other body systems (p. 222)

strep throat   a disease that causes a sore throat and a fever (p. 102)

stress   emotional pressure people feel when they face a difficulty (p. 201)

stressor   an event that causes stress (p. 201)

stress response   an automatic reaction to a feeling of stress (p. 201)

synapse   the space between neurons (p. 25)

syphilis   a sexually transmitted disease that attacks many parts of the body (p. 105)

T

temperament   the emotional nature of a person (p. 168)

tendon   a type of tissue that connects muscles to bones or other muscles (p. 12)

testes the male organs that make sperm cells and hormones; singular is *testis* (p. 75)

testosterone the male sex hormone (p. 75)

tissues groups of cells that work together (p. 3)

trachea a tube that brings air from your throat into your lungs; also called the windpipe (p. 50)

tranquilizer a drug that has a calming effect (p. 231)

transplant to remove an organ, such as the heart, and place it in another person's body (p. 6)

triceps the muscle in the back of the upper arm (p. 13)

trimesters the three periods a pregnancy is often divided into (p. 79)

truck drivers a slang term for *amphetamines* (p. 232)

tuberculosis a very contagious disease that affects the lungs; also called TB (p. 102)

tumor a clump of tissue caused when abnormal cells grow together (p. 106)

U

umbilical cord a tube that connects the baby to the placenta (p. 78)

unsaturated fats fats that are easier for the body to break down (p. 143)

uppers a slang term for *amphetamines* (p. 232)

ureters the tubes that carry urine leaving the kidneys (p. 65)

urethra the tube through which the urine leaves the body (p. 65)

urinary bladder the organ that collects and stores urine until it leaves the body (p. 65)

urine a mix of water and wastes (p. 65)

uterus the female organ in which a fertilized egg develops into a baby (p. 76)

V

vaccine a substance that stimulates immunity to a disease (pp. 49, 214)

vagina a canal that leads from a woman's uterus to the outside of her body (p. 76)

vein a blood vessel that carries blood toward the heart (p. 41)

ventricles the two bottom chambers of the heart (p. 40)

vertigo a feeling of dizziness (p. 32)

villi tiny finger-shaped structures in the small intestine that help the body absorb nutrients from food (p. 61)

virus a microscopic structure that can cause disease (p. 93)

visualizations positive thinking used to control stress (p. 205)

vitamin a nutrient found in foods that the body needs and that is made by other organisms (p. 146)

voluntary muscle a muscle that moves with your control (p. 12)

voluntary nerve action a response that happens when you think about doing something first (p. 27)

W

weed a slang term for *marijuana* (p. 235)

white blood cell a type of cell that helps the body fight infection (p. 42)

whole body scans a costly scanning technology that looks at the whole body for disease (p. 111)

withdrawal a physical reaction to the removal of an addictive substance in the body (p. 223)

Z

zero-emission vehicles (ZEVs) cars that do not release harmful pollutants into the air (p. 294)

Index

A

AA (Alcoholics Anonymous), 221
Abstain from alcohol, 220
Abstinence, sexual, 105, 265, 270
    practicing, 270
Abuse, drug, *See* drug abuse
Accidents, preventing, 115–125
Acid, 234
Acid reflux, 62
Acne, 129, 134
    preventing, 135
Acquired immune deficiency syndrome, *See* AIDS
Action plan, SHL6, 261
Acupuncture, 204
Addiction, 213, 219
    alcohol, 219–221
    tobacco, 223
Addictive, 229, 231
Adolescence, 253
Adrenaline, 201
Advertising, sexuality and, 267
Advocacy group, 291, 298
Aggressive, communication style, 277, 279
AIDS (Acquired Immune Deficiency Syndrome), 92, 101, 103–104, 268–269, 307
    misconceptions about, 104
Air pollution, 294
    indoor, accessing information about, 299
Alcohol, 28, 67, 125
    body and, 216
    motor vehicle safety and, 218
Alcohol addiction, 219–221
Alcoholics, 213, 219
    children of, 220
    families of, 219
    help for, 221
Alcoholics Anonymous (AA), 221
Alcoholism, 213, 219
    teenagers and, 220
Alternative medicine, 204

Alveoli, 51
Alzheimer's disease, 159, 162
American Red Cross, 297
Amino acids, 143
Amniotic sac, 79
Amphetamines, 229, 232
Anemia, 45
Angel dust, 235
Anger, 175, 185
Anorexia, 187, 195–196
Antibiotics, 89, 91, 214
Antibodies, 48, 95
Antiperspirants, 129, 134
Antiseptic, 90
Antisocial personality disorder, 191
Anxiety, 187
Appendix, 58
Arteries, 41
Arteriogram, 111
Arteriosclerosis, 108
Arthritis, 3, 10
Artificial joints, 11
Assertive, communication style, 277, 279
Asthma, 52, 101, 109, 307
Atherosclerosis, 108
Athlete's foot, 134
Atria, 40
Automatic impulses, 26
Axons, 21, 25

B

Back muscles, 16
Bacteria, 89, 93
Bacterial diseases, 93
Bad breath, 133
BAL (blood alcohol level), 213, 217
Balanced diet, 142
Ball-and-socket joints, 7
Barbero, Michael, 247
Barbiturates, 229, 231
Behavior, school social groups and, 286
Benign tumors, 106
Biceps muscle, 13
Bicycle safety, 122

Bile, 57, 60
Bipolar disorder, 191
Blackheads, 134
Blind people, 35
Blood, 42
Blood alcohol level (BAL), 213, 217
Blood donation, 43
Blood transfusions, 104
Blood types, 42
Blood vessels, 41
Body
    alcohol and, 216
    mind and, SHL2, 166
    stress and, SHL2, 201–202
    tobacco and, 222–223
Body language, 277, 280
    positive, 281
Body systems, 3
    circulatory, immune, and respiratory systems, 39–53
    digestive and excretory systems, 57–69
    nervous and endocrine systems, 21–35
    reproductive system, 73–83
    skeletal and muscular systems, 3–17
Bone marrow, 3, 6
Bones, 4–10
    broken, 8
    structure of, 5–6
Braces on teeth, 133
Brain, 22–23, 159–161
    development of, 161
    parts of, 23, 160
Brain damage, 161
Brain imaging technology, 162
Brain injuries, 171
Brainstem, 22, 23
Brain tumors, 161
Breath Analyzers, 217
Broken bones, 8
Bronchi, 50
Bronchitis, 223
Bulimia, 187, 195–196
Burns, Joe, 181

Ultrasound, prenatal, 111
Umbilical cord, 78
Unsaturated fats, 143
Uppers, 232
Ureters, 65
Urethra, 65
Urinary bladder, 65
Urine, 65
Uterus, 73, 76

V

Vaccines, 49, 95, 214
Vagina, 73, 76
Values, 260
Varicose veins, 45
Veins, 41
Venereal warts, 307
Ventricles, 40
Vertigo, 32
Violence
    drug abuse and, 237
    preventing, 292

Viral diseases, 93
Viruses, 89, 93
Visualizations, 201, 205
Vitamin A, 146, 305
Vitamin B_1, 146, 305
Vitamin B_2, 305
Vitamin B_{12}, 305
Vitamin C, 102, 146, 305
Vitamin D, 10, 146, 305
Vitamin E, 305
Vitamin K, 305
Vitamins, 141, 146, 305
Voluntary muscles, 3, 12
Voluntary nerve action, 27
Volunteer in community, 295
Vomiting, 62

W

Walking, safe, 124
Water, 142
    safety in, 120
Water pollution, 294

Weed, 235
Weight management, 152
White blood cells, 42, 48
Whiteheads, 134
WHO (World Health
    Organization), 296
Whole body scans, 111
Wisdom teeth, 132
Withdrawal, 213, 223
World Health Organization
    (WHO), 296
Worry, 175

Y

Yeast infection, 307

Z

Zero-emission vehicles, 294

Acknowledgments

Illustrations
All illustrations © Pearson Learning Group unless otherwise noted.
Page 134: Argosy.
Photographs
All photographs © Pearson Learning Group unless otherwise noted.
Cover: *t.l.* © Michael Matisse/PhotoDisc, Inc., *m.t.* © Digital Vision/Getty Images, *t.r.* © C Squared Studios/PhotoDisc, Inc., *b.l.* © SuperStock, Inc., *m.b.* © C Squared Studios/PhotoDisc, Inc., Page SHL2: © BananaStock/Fotosearch.com, SHL3: © Michael Wong/Corbis, SHL4: David Young-Wolff/PhotoEdit, SHL6: © Spencer Grant/PhotoEdit, SHL7: Department of Health and Human Services, SHL8: © Jeff Greenberg/PhotoEdit, 2: © Frank Siteman/Omni-Photo Communications, Inc., 5: © Carolina Biological/Visuals Unlimited, 7: © Bill Hickey/The Image Bank/Getty Images, 11: © Zephyr/Photo Researchers, Inc., 12: © Quest/Science Photo Library/Photo Researchers, Inc., 14: *t.* © Eyewire Collection/Getty Images, Inc., *b.* © Hekimian Julien/Corbis Sygma, 20: © Alfred Pasieka/Science Photo Library/Photo Researchers, Inc., 24: Courtesy of Tim O'Brien, 32: © Prentice Hall, Inc., 35: © Fritz Prenzel/Animals Animals/Earth Scenes, 38: © Oliver Meckes/Photo Researchers, Inc., 43: © AP/Wide World Photo, 44: © Science Photo Library/Photo Researchers, Inc., 48: © Manfred Kage/Peter Arnold, Inc., 49: *t.* © Michael Newman/PhotoEdit, 49: *b.* © Vo Trung Dung/Corbis Sygma, 56: © ISM/Phototake, 61: © R.A. Becker/Custom Medical Stock Photo, 63: © AP/Wide World Photo, 66: © Alfred Pasieka/Photo Researchers, Inc., 68: © Bob Daemmrich/The Image Works Incorporated, 69: © David Young-Wolff/PhotoEdit, 72: © David M. Phillips/The Population Council/Photo Researchers, Inc., 78: Dr. Y. Nikas/Phototake, 80: © Tek Image/Photo Researchers, Inc., 88: © A. David Hazy/Custom Medical Stock Photo, 90: © Mary Kate Denny/PhotoEdit, 91: © Bettmann/Corbis, 92: © A.B. Dowsett/Science Photo Library/Photo Researchers, Inc., 96: © Bettmann/Corbis, 97: © Getty Images, 100: © Photo Researchers, Inc., 102: © David M. Phillips/Photo Researchers, Inc., 104: © John Henley/Corbis, 106: © Dr. P. Marazzi/Photo Researchers, Inc., 107: Courtesy of Astrazeneca, 109: © Michael Newman/PhotoEdit, 110: © Oliver Meckes & Nicole Ottowa/Photo Researchers, Inc., 111: © Biophoto Associates/Photo Researchers, Inc., 114: © PhotoDisc, Inc., 117: © Tony Freeman/PhotoEdit, 120: © Angela Wyant/Stone/Getty Images, 122: © Jeff Greenberg/PhotoEdit, 123: © David Young-Wolff/PhotoEdit, 128: © SW Productions/PhotoDisc, Inc., 130: © Ryan McVay/PhotoDisc, Inc., 131: © Richard T. Nowitz/Phototake, 133: © George Disario/Corbis, 135: © SW Productions/PhotoDisc, Inc., 140: © Michelle D. Bridwell/PhotoEdit, 144: © James Jackson/Stone/Getty Images, 149: © Tracy Frankel/The Image Bank/Getty Images, 153: © Lori Adamski Peek/Stone/Getty Images, 158: © Steve Skjold/PhotoEdit, 162: *t.* © NIH/Science Source/Photo Researchers, Inc., *b.* © Richard Price/Taxi/Getty Images, 163: © Bob Daemmrich/PhotoEdit, 166: © Richard Meats/Stone/Getty Images, 167: © Ryan McVay/PhotoDisc, Inc., 169: © Spencer Grant/PhotoEdit, 172 © E. Dygas/Taxi/Getty Images, 176: © Mary Kate Denny/PhotoEdit, 177: © Ed Bock/Corbis, 178: © Dana White/PhotoEdit, 181: © Tony Freeman/PhotoEdit, 186: © Dennis McDonald/PhotoEdit, 188: © Myrleen Ferguson Cate/PhotoEdit, 189: © George Hall/Corbis, 194: © Mary Kate Denny/PhotoEdit, 196: © Michael Newman/PhotoEdit, 200: © Marc Romanelli/The Image Bank/Getty Images, 203: © Jose Luis Pelaez/Corbis, 204: © Bob Daemmrich/The Image Works Incorporated, 206: © Laurent/Photo Researchers, Inc., 207: © Ian Shaw/Stone/Getty Images, 212: © AP/Wide World Photo, 214: © George B. Diebold/Corbis, 216: © DK Images, 217: © Custom Medical Stock Photo, Inc., 218: © Bill Beatty/Visuals Unlimited, 220: © SW Productions/PhotoDisc, Inc., 222: © James Stevenson/Photo Researchers, Inc., 224: © Tony Freeman/PhotoEdit, 228: © David Young-Wolff/Stone/Getty Images, 232–233: Drug Enforcement Administration/U.S. Department of Justice, 236: © Charles Gupton/Corbis, 239: © Don Smetzer/PhotoEdit, 244: © Rob Lewine/Corbis, 246: © Jeff Greenberg/PhotoEdit, 247: © Tom McCarthy/PhotoEdit, 249: © David Young-Wolff/PhotoEdit, 252: © Tom Rosenthal/SuperStock, Inc., 254: © Ariel Skelley/Corbis, 256: © David Young-Wolff/Stone/Getty Images, 257: © Michael Newman/PhotoEdit, 259: © Ariel Skelley/Corbis, 261: © David Young-Wolff/PhotoEdit, 264: © Patrick Molnar/Taxi/Getty Images, 266: © David R. Frazier/David R. Frazier Photography 267: © Aline Maurice/Taxi/Getty Images, 269: © Don Johnston/Stone/Getty Images, 271: © Adamsmith/Taxi/Getty Images, 276: © Phil Martin/PhotoEdit, 278: © Michael Newman/PhotoEdit, 283: © Bruce Ayres/Stone/Getty Images, 286: © Spencer Grant/PhotoEdit, 290: © Johnathan Nourak/PhotoEdit, 293: © Bob Coates/Index Stock Imagery, Inc., 294: © Getty Images, 295: © Myrleen Ferguson Cate/PhotoEdit, 297: © Getty Images, 299: © Photo Researchers, Inc.